RED EARTH, WHITE LIES

Other Books by Vine Deloria, Jr.

Custer Died for Your Sins
We Talk, You Listen
Of Utmost Good Faith
God Is Red
Behind the Trail of Broken Treaties
The Indian Affair
Indians of the Pacific Northwest
The Metaphysics of Modern Existence
American Indians, American Justice (with Clifford Lytle)
A Sender of Words (edited)
The Nations Within (with Clifford Lytle)
The Aggressions of Civilization (edited with Sandra Cadwalader)
American Indian Policy in the Twentieth Century (edited)
Frank Waters, Man and Mystic (edited)

RED EARTH, WHITE LIES

Native Americans and
the Myth of Scientific Fact

Vine Deloria, Jr.

FULCRUM PUBLISHING

GOLDEN, COLORADO

Library of Congress Cataloging-in-Publication Data
Deloria, Vine.
Red earth, white lies : Native Americans and the myth of
scientific fact / Vine Deloria, Jr
p.   cm.
Originally published: New York : Scribner, c1995
Includes bibliographical references and index.
ISBN 1-55591-388-1 (pbk.)
1. Indians of North America—Folklore. 2. Indian philosophy—
North America. 3. Oral tradition—North America—History and
criticism. 4. Science—Philosophy. 5. Religion and science
6. Human evolution—Religious aspects—Christianity. I. Title
E98.F6D35   1997
398'.08997—dc21   97–21689
                                        CIP

Printed in United States of America

0 9 8 7 6 5

FULCRUM PUBLISHING
16100 Table Mountain Parkway, Suite 300
Golden, Colorado 80403
(800) 992-2908 ▪ (303) 277-1623

Dedicated to
my father, Vine Deloria, Sr.,
who told me many stories.

Contents

Preface

OTHER THAN *Custer Died for Your Sins,* this book has been the most pleasant to write and the most fun to defend. Learning that I did not believe in the Bering Strait theory, the anthropology department at Colorado University, in a series of secretive e-mails, decided I was a racist reactionary trying to destroy their fictional enterprise and agreed not to invite me to speak to them. Had they tendered an invitation, I doubt if I would have appeared, so perhaps a point was scored on each side. Around the country the reception has pleased me to no end. Instead of defending me, many Indian students decided to call my bluff and went to the libraries and found I was right—no good evidence except the mental illness of the academy exists supporting this theory. Across the board, younger professors and graduate students approved of the book, and the old guard formed militia movements to protect the tottering bastions of Western knowledge. Most important, however, is the flood of new articles and newspaper clippings people from all over the world have sent me—almost all supporting the ideas in this book.

I have changed some of the chapters a bit to include new materials, primarily adding a short section on Alaska. It seems that scholars studying Alaska Natives have given the oral tradition its just due—at least for the matching of recent events of geological importance with stories preserved by the people. If that same spirit were shown by scholars working with tribes of the lower forty-eight states, we would have some real change in the way we look at the history of North America.

Scholars working with the extinction of the Pleistocene megafauna continue to pass their personal fantasies off as science while becoming more absurd with each effort to come to grips with the problem. Instead of dealing with the millions of carcasses in Alaska and Siberia that suggest a massive planet-wide catastrophe, as I suggest here, they have now concocted a theory that the legendary Paleo-Indians brought infectious diseases with them across the Bering Strait. These diseases, according to alleged scholars, were carried by "frogs, rats, birds, parasites, and other living baggage" that accompanied the Indians. As a rule, species do not give diseases to other species, except when scholars require them to do so to bolster a theory that has no evidence to support it. After the book came out I had occasion to visit Rancho La Brea and stood in front of a 12-foot mammoth skeleton. I would have had to get right under the beast to even touch it with a spear. I suggest that the scholars advocating the "blitzkrieg" actually go see how big these creatures actually were.

In February 1977, a group of the most reactionary scholars issued a report from Monte Verde, Chile, to the effect that they now accepted a new date for the occupation of the Western Hemisphere by Paleo-Indians. According to newspaper reports, a site was unanimously dated at 12,500 years ago. The report then noted that "the ancestors of those Chilean settlers somehow managed to travel some 10,000 miles from the Bering Strait to southern South America in only a few hundred years." In other words, another anthropological sleight of hand—hitting the Alaskan shore, these fictional people made a beeline for South America, forsaking the green pastures of North America, and trooping through almost impassable deserts and jungles to arrive in time to save anthropological theories 12,500 years later. The admission was hailed as a major event in anthropology and archaeology. The headline should have read "Scholars Moving Reluctantly Toward Sobreity."

At any rate, enjoy the book. Watch the newspapers for more startling admissions that all is not right in Western Hemisphere prehistory and ask your local scholar to provide evidence for the fantastic scenarios that are being passed off as "science." You will enjoy watching them squirm and change the subject.

Introduction

LIKE ALMOST EVERYONE ELSE IN AMERICA, I grew up believing the myth of the objective scientist. Fortunately I was raised on the edges of two very distinct cultures, western European and American Indian—the great Sioux tribe to be exact. Growing up in a little border town on the Pine Ridge Indian Reservation in South Dakota, we all knew that we could never understand the complicated theories of science, literature, and philosophy that were common knowledge among sophisticated people in the cities. So we mostly didn't try, simply believing that somewhere all the contradictions were resolved satisfactorily—at least in the minds of those more intelligent than we were.

As time passed I became an avid reader of popular scientific books, wanting to know as much as I could about the world in which I lived. Gradually I began to see a pattern of nonsense in much scientific writing. Scientific explanations given regarding the origins or functioning of various phenomena simply didn't make sense. Bowing to scientific authority, I kept to the premise that brighter people than I understood the complexities of nature and I assumed that no real contradictions existed. Then one day while reading Jacob Bronowski's *The Ascent of Man*, recommended to me by a bevy of noted anthropologists, I came across the following sentence:

> Why are the Lapps white? Man began with a dark skin; the sunlight makes vitamin D in his skin, and if he had been white in Africa, it would make too much. But in the north, man needs to let in all the sunlight there is to make enough vitamin D, and natural selection therefore favoured those with whiter skins.[1]

I had encountered the same idea many times before in the publications of a number of prestigious scientific writers, but until then it never struck me as odd. The fact is that Lapps may have whiter skins than Africans, but they do not run around naked to absorb the sunlight's vitamin D. Indeed, it is the Africans who are often bare in the tropical sun. The Lapps are always heavily clothed to protect themselves from the cold. Whatever "natural selection" did, skin color obviously played no part.

After that, I had difficulty taking scientific doctrines seriously and I began to make notes of the more sublime authoritative statements I found in scientific writing to remind myself of its essential foolishness. As my faith in science decreased geometrically over the years, like many former acolytes, I was embarrassed by my former allegiance. But I did not think that scientific doctrines were harmful. Then I began to hear how my ancestors had ruthlessly slaughtered the Pleistocene megafauna and I began to read about this hypothesis. As I saw rednecks and conservative newspaper columnists rant and rave over the supposed destruction of these large animals, I saw a determined effort to smear American Indians as being worse ecologists than our present industrialists. Thus, I decided to write this book, offering an alternative explanation for the demise of the great animals.

Beginning in 1992 the American Indian Science and Engineering Society sponsored a series of conferences on traditional knowledge of Indian tribes. We started to build a network of knowledgeable traditional people who brought together the beginnings of a body of information that sheds light on the pre-Columbian history of this hemisphere and perhaps even the history of our planet. This knowledge describes unusual events and often gives a reasonable alternative explanation of how things happened. It is the distilled memories of thousands of years of living in North America. I do believe that perhaps only 10 percent of the information that Indians possess is presently in print and available for discussion. And I hope that in the years ahead even more data will come to light or be made available by our elders.

I have a rule when quoting elders and traditional leaders in a book. Basically the stories belong to them, and so I do not want to be the first person to put a story in print. I therefore try to find a published account that confirms what an elder has told me and use that version instead of the tradition as related to me by the elder. Thus, there are a considerable number of stories that I could have put into this book that will not appear until the elders themselves authorize them to be published.

But I do hope that this book encourages elders to give us some of their knowledge before they pass on. I also hope I have given enough direction in this book so that the next generation of Indians will respect, cherish, and rescue the remaining bits of information that our people possess. With the collision of the Shoemaker comet into Jupiter, the era of uniformitarian orthodoxy must come to an end. Minds that have been closed for nearly half a millennium can now be opened to see what really has happened to our planet in the past—and that past is not as distant as we might suppose.

1

Behind the
Buckskin Curtain

WHEN INDIAN BINGO GAMES ARE HUMMING in almost every nook and cranny of our land, stealing the most sacred ritual of the Roman Catholic Church and gathering the white man's coin as quickly as it can reasonably be retrieved, progress is being made. When multitudes of young whites roam the West convinced they are Oglala Sioux Pipe Carriers and on a holy mission to protect "Mother Earth," and when priests and ministers, scientists and drug companies, ecologists and environmentalists are crowding the reservations in search of new rituals, new medicines, or new ideas about the land, it would appear as if American Indians finally have it made. Indeed, some tribal chairmen are now well-heeled Republicans worried about gun control, moral fiber, and prayer in schools. In many respects American Indians are looking increasingly like middle-class Americans.

Beginning in 1960, the federal census allowed people to self-identify their ethnic or racial background, and in the past three decades a startling jump in the Indian population has occurred. Where there were over half a million Indians in the United States in 1960, in the last census the Cherokees alone totaled over 360,000, primarily the result of consciousness-raising efforts of New Age enthusiasts but nevertheless welcome as a politically significant figure. As whites get more familiar with Indian symbols and beliefs we can expect both the national figures and the Cherokee figures to skyrocket beyond belief by the year 2000. Indeed, today it is popular to be an Indian. Within a decade it may be a necessity. People are not going to want to take the blame for the sorry state of the nation, and claiming allegiance with the most helpless racial minority may well be the way to escape accusations.

These positive symbols of prosperous buckskin are not the whole story, unfortunately. Nothing is calm beneath the veneer of Indian country, and it may be that we are seeing the final absorption of the original inhabitants in the modern consumer society. The push for education in the last generation has done more to erode the sense of Indian identity than any integration program the government previously attempted. The irony of the situation is that Indians truly believed that by seeking a better life for their children through education, much could be accomplished. College and graduate education, however, have now created a generation of technicians and professionals who also happen to have Indian blood. People want the good life and they are prepared to throw away their past in order to get it.

The situation would be perilous indeed were it not for the fact that the white majority spent the past generation tearing down its culture also. A winsome essay in *Time* magazine during the summer of 1994 asked the question of how to define things once everything in society is "hip." The problem of being "hip" plays right into our hands. Indians can always become whites because the requirements are not very rigorous, but can whites really become Indians? A good many people seriously want to know. They are discontented with their society, their government, their religion, and everything around them and nothing is more appealing than to cast aside all inhibitions and stride back into the wilderness, or at least a wilderness theme park, seeking the nobility of the wily savage who once physically fought civilization and now, symbolically at least, is prepared to do it again.

Three areas exist that contain tremendous barriers to any effort of whites to become Indians. These areas, unless they are given careful and serious attention by the next generation of Indians, may prove fatal to Indian efforts to remain faithful to whatever traditions are still being practiced. While it may appear that Indians are adopting the values and practices of American culture, in science, in religion, and in forms of social interaction—most prominently in government—there is still a tremendous gap between the beliefs and the practices of both whites and Indians.

These three areas of conflict and misunderstanding were present at the beginning of colonial discovery days; they have defined the terms of the conflict between the indigenous peoples and the invaders for more than five hundred years and they remain potent, because they provide the definition of what civilized society should be. Our present view of government, our avoidance of allegiance to high spiritual powers, and our exclusively scientific understanding of our world will continue to guide our thoughts and activities in the future and bring us to a complete collapse unless we achieve more mature understanding of our planet, its history, and the rest of the universe. Much of Western science must go, all of Western religion should go, and if we are in any way successful in ridding ourselves of these burdens, we will find that we can fundamentally change government so that it will function more sensibly and enable us to solve our problems.

Science and religion are inherited ways of believing certain things about the world. A good many of our problems today are a result of the perpetuation of dreadfully outmoded beliefs derived from the Near Eastern/European past that do not correspond to what our science is discovering today or to the remembered experiences of non-Western peoples across the globe. Even the purest forms of scientific and religious expression are rooted in the unconscious metaphysics of the past, and critical examination of the roots of the basic doctrines in these areas will reveal the inadequacy of our beliefs. Government is simply the way we organize ourselves and move masses of people to behave in certain ways. Government we cannot do without, but we certainly do not have to continue to act as if it were *only* a way of moving masses of people and manipulating their beliefs and behavior.

When Europeans arrived on these shores they brought with them a powerful technology. For much of the first four hundred years of contact, technology dealt Indians the hardest blows. Mechanical devices from the musket to the iron kettle to the railroad made it a certainty that Indians would lose the military battle to maintain their independence. Technology made it certain that no tribe would be able to maintain its beliefs in the spiritual world when it was apparent that whites had breached certain fundamental ways of living in that spiritual world and in this breach had foreclosed even the wisest of their people from understanding the larger arena in which human destiny was being played out. Whites had already traded spiritual insight for material comfort, and once trade of material things came to characterize the Indian relationship with whites, Indians lost much of their spiritual heritage also.

In many ways technology serves us and makes our lives better. Behind and beneath technology, however, in scientific theories and doctrines, lurk a large number of misperceptions, badly directed emphases, and unresolved philosophical problems. As Western civilization grew and took dominance over the world, it failed to resolve some basic issues. A view of the natural world as primarily physical matter with little spiritual content took hold and became the practical metaphysics for human affairs. During the European Middle Ages a basic split in perspective occurred when *reason* and *revelation*, the twin paths for finding truth in the minds of Western thinkers, were divided into sacred and secular and became equivalent but independent bodies of knowledge. Once reason became independent, its only referent point was the human mind and in particular the middle-class, educated, European mind. Every society needs educated people, but the primary responsibility of educated people is to bring wisdom back into the community and make it available to others so that the lives they are leading make sense.

European thinkers did not perform their proper social function. Science and philosophy simply copied the institutional paths already taken by Western religion and mystified themselves so that one of the maxims of recent Western civilization has been to declare something to be "academic"—meaning that intelligent solutions to problems are in fact illusory because they are devised by people sheltered from the realities of daily life. Western people prefer political solutions, rejecting the principles involved and choosing a practical, often compromising resolution of the problem. This tradition, as we experience it today, is the tendency to authorize or fund an amazing number of studies whose conclusions are ultimately rejected so that we can devise a political solution that will enable us to avoid understanding or confronting the issue altogether and which usually is satisfactory to no one.

Institutionalization of science took many forms: the increasing tendency of people to look to scientists for reliable explanations about the world, the development of universities and colleges, sponsorship of scientific research by wealthy patrons and eventually the state. Most of all, however, it meant that scientists would come to act like priests and defer to doctrine and dogma when determining what truths would be admitted, how they would be phrased, and how scientists themselves would be protected from the questions of the mass of people whose lives were becoming increasingly dependent on them. In our society we have been trained to believe that scientists search for, examine, and articulate truths about the natural world and about ourselves. They don't. But they do search for, take captive, and protect the social and economic status of scientists. As many lies are told to protect scientific doctrine as were ever told to protect "the church."

This condition might have corrected itself had not Western science given status to the winter ramblings of René Descartes, who mistakenly concluded that mind and matter were independent things. The idea had already been an acceptable proposition in Europe theologically thanks to the Inquisition, which sought to save the soul by destroying the body. In its more popular use we must remember the words of Arnaud, the papal legate at the siege of Béziers in the Albigensian Crusade against the Cathars of southern France. When asked if Catholics should be spared from the massacre of the city's inhabitants, Arnaud answered, "Kill them all, for God knows His Own."[1] The Cartesian bifurcation of nature was the more fatal because it encouraged succeeding generations of scientists to treat an obviously living universe as if it were an inert object. Eventually this idea meant useless experimentation on other forms of life as if they had no sentient capabilities at all. When the Western slaveowners and military encountered darker-skinned people, this doctrine enabled atrocities of unimagined proportions to be administered without accumulating much guilt along the way.

American Indians encountered Western science in its most deadly technological expression in the Gatling guns at Wounded Knee, but already forces nearly as sinister lurked in the scientific mind. Early speculations about the origin of Indians relied on the biblical accounts of human history, and consequently the Indians were often seen as descendants of ancient Hebrews, the question being merely to properly identify which remnant of mankind had become Indians—the peoples after the great flood of Noah or fragments of the Jews who survived the Babylonian Captivity. It was believed that Indians could not travel by water, and so the Bering Strait became first the ecclesiastical and then the scientific trail from Jerusalem to North America.

Then came the idea that human cranial capacity demonstrated the intelligence of the different races, a misbegotten but piously proclaimed scientific truth. Indians were hardly on their reservations before government employees began robbing graves at night to sever skulls from freshly buried bodies for eastern scientists to measure in an attempt to prove a wholly spurious scientific theory. Indeed, it may have been that Indians were unnecessarily slaughtered in battles, since it was a custom to simply ship bodies of Indians killed by the army to eastern laboratories for use in various experiments. Some Eskimos staying at a New York museum to help the scientists died and were boiled down for further skeletal use instead of receiving a decent burial. Even today, dark rumors continue to circulate concerning the use of Indians by the Indian

Health Service to test experimental drugs. Some years ago there were real questions concerning the number of Indian women being sterilized at government clinics without their knowledge or consent.

These practical examples of science gone mad are but minor points in the long run. More important for our purposes, while not forgetting the horrors of some scientific behavior, is the impact of scientific doctrine on the status of Indians in American society. Regardless of what Indians have said concerning their origins, their migrations, their experiences with birds, animals, lands, waters, mountains, and other peoples, the scientists have maintained a stranglehold on the definitions of what respectable and reliable human experiences are. The Indian explanation is always cast aside as a superstition, precluding Indians from having an acceptable status as human beings, and reducing them in the eyes of educated people to a prehuman level of ignorance. Indians must simply take whatever status they have been granted by scientists at that point at which they have become acceptable to science. There was a terrible reaction in 1969 when I accused anthropologists of treating Indians like scientific specimens. Now, after a quarter of a century, Indians are no longer "informants"; they are now seen quite often as colleagues.

The stereotypical image of American Indians as childlike, superstitious creatures still remains in the popular American mind—a subhuman species that really has no feelings, values, or inherent worth. This attitude permeates American society because Americans have been taught that "scientists" are always right, that they have no personal biases, and that they do not lie, three fictions that are impossible to defeat. A current example will suffice. For several decades Indians have complained about the use of grotesque cartoon figures of Indians used as sports mascots, the most prominent being that used by the Washington professional football team. We have been lectured by every redneck peckerwood who can man a typewriter about how harmless these names and symbols are. Some columnists have even written columns warning that if the Indians are successful, there will be protests by Lions, Tigers, and Marlins—indicating that Indians are still classified as wildlife instead of human beings. Indeed, in the 1960s a popular children's publisher published a little book on animals and their babies and featured an Indian woman and a baby right along with the mother bear and her cubs and other mother-baby animal figures. Scientists may not have intended to portray Indians as animals rather than humans, but their insistence that Indians are outside the mainstream of human experience produces precisely these reactions in the public mind. Yet there is a tremendous schizophrenia among non-Indians regarding the Indian image, since so many people want to claim Indian blood. It is the nobility and authenticity of nature that they see in Indians and want—they want to be pure and natural.

In recent years, we have seen a number of famous personalities in the field of sports, particularly sports broadcasting, fall into disgrace because of their racial remarks about other minorities—primarily African Americans and Jews. Thus a Dodger executive is fired because he casually stated that African Americans were not smart enough to work in the front office; Jimmy the Greek related on a national telecast that African Americans had large muscles with their hips apparently extending farther up the back than whites because slaveowners had bred them for these characteristics. Both of these personalities were fired for the better image of the game. Marge Schott, owner of the Cincinnati Reds, in *private* conversation, made derogatory remarks about African Americans and Jews and received a year's suspension from baseball. Why are these remarks regarded as mortal sins even while derogatory symbols and names of American Indians spark the opposite effect and trigger intense chastisement of Indians when protested?

It is not enough to point out that African-American ballplayers make a lot of money for the club owners or to suggest that black Americans have greater political clout than American Indians. Most non-Indians are happy and zealous to claim a Cherokee grandmother in their family tree, whereas we hear very little about

African-American blood in the veins of otherwise white people. The constant drumbeat of scientific personalities who manipulate the public's image of Indians by describing archaeological horizons instead of societies, speaking of hunter-gatherers instead of communities, and attacking Indian knowledge of the past as fictional mythology, has created a situation in which the average citizen is greatly surprised to learn that Indians are offended by racial slurs and insults. In this book we will examine some of the scientific doctrines that contribute to this callous treatment.

Religion, in any usual meaning of the term, ceased to exist in America long ago. Indeed, any higher deity exists for Americans only insofar as he or she can guarantee great sex, lots of money, social prestige (read celebrity), a winning football team, and someone to hate. American denominations are busy frantically trying to appeal to the unchurched by redefining sin so that anyone who has any quirks at all can still be a regular part of the congregation, if not the ministry. It is impossible to sin today because all of the really good sins have become Christian virtues, much as Greed, once one of the Seven Deadly Sins, has become the chief American and Republican virtue.

In spite of the overtures made in recent years by Christian denominations toward traditional Indian religions and practices—such as bishops wearing warbonnets at services, pipes and other traditional objects used to bless congregations, and occasional prayers for the Earth—one fundamental facet of Christianity must always detour any effort to come to grips with reality. Christianity was not designed to explain anything about this planet or the meaning of human life.

Every single ceremonial act of the Christian tradition is based on the belief that history was coming to an end, that the believers would be taken up to heaven, a place radically different from this Earth, and that everything about the Earth—and about human society and societies—was evil. Since the expectations of a dramatic judgment day have not been fulfilled, and since Western technology has done a pretty decent job of creating heaven here on Earth, Christian preachers can no longer instill fear into people, and the traditional ceremonies have become empty forms, creating the need for most churches to borrow from American entertainment media in order to fill their pews.

Christianity has been the curse of all cultures into which it has intruded. It has offered eternal life somewhere else and produced social and individual disintegration. Even today its chief personalities fall one after another into disrepute. Catholic priests prey on their parishioners; televangelists engage in fraudulent financial practices or are seen in the seedy parts of town on sexual escapades. Clergy extol the virtues of "the church" but rarely speak of God, and today we have large numbers of them desperately trying to get into Indian ceremonies to experience "spirituality."

Throughout American history we have all suffered because of the European religious heritage. Shortly after the discovery of the Western Hemisphere the Protestant Reformation began, initiating a period of intense religious bigotry and intolerance that crested with the Spanish Inquisition and the Salem witch hunts. Systematic torture of unbelievable pain was visited upon anyone suspected of harboring heretical thoughts. And who could tell what would be heretical thoughts when often the religious choice of the monarch determined the official religion of the state?

Americans initially did not do a whole lot better than their parents and grandparents had done while in Europe. Quakers and Catholic priests could be killed if, after they had been expelled from Massachusetts, they were again found there. A number of decent parishioners were killed at Salem primarily for being old. On the whole, however, Americans did reasonably well in holding their religious savagery in check. A priest once told me that things would have been much worse if the people who did all these things were not Christians. I remain skeptical. Virginia, where all the decent intellectuals except Ben Franklin seem to have lived, set an example of religious freedom primarily because of Thomas Jefferson, a deist, not a Christian.

With the ratification of the Constitution, Americans insisted on the inclusion of the Bill of Rights so that they could postpone the time when government would become their oppressor until the twentieth century and so that they could possess guns. At least Republican congressional candidates have insisted on these motives as being foremost in the minds of the constitutional fathers. Included in the First Amendment was an effort to deal with the European religious heritage. Congress was prohibited from making any "law respecting the establishment of religion, or prohibiting the free exercise thereof." Like almost every other phrase in the Constitution, the religious clauses of the First Amendment have come to mean many things in our society.

Looking over all the case law that has accumulated in regard to these prohibitions, we see two major trends emerge: we can always devise the proper rhetoric so that church schools can be funded as if they were public schools, and we can prohibit the display of almost anything hinting at religious belief or sentiment in any public place.

American Indians were not included in the Constitution, since they were regarded as enemies on the frontier who might suddenly align with England or France, perhaps even Spain, and prove to be the critical mass that would snuff out the independence of the United States. As the tribes grew weaker militarily and lost their political and economic independence, the government and the churches quickly aligned. Soon, in spite of constitutional prohibitions, Christianity was made the official religion of Indian reservations, and traditional tribal religions were banned. Indeed, Indian students were allowed to read in their own language only if the Bible had been printed in it. In 1870, President Grant simply handed out religious monopolies to the respective denominations in different parts of the country. Thus some tribes are Lutherans, Catholics, Episcopalians, and Presbyterians not because a missionary eloquently convinced them of the validity of Christianity but because the Grant administration assigned tribes to particular denominations.

In 1934, Indians were finally granted a measure of religious freedom on their own lands by repeal of some of the more punitive measures of the Interior Department and by the granting of self-government. Since that time, in varying degrees, traditional people have begun to feel safer in performing ceremonies within the reservation and sometimes even at locations outside the reservation. In 1988 and 1990, the Supreme Court of the United States chilled Indian religious practices by taking two cases that need not have been heard and writing opinions so bizarre that they disrupted the case law of the First Amendment to the detriment of all religious bodies.[2] Previous case law had laboriously constructed what was called the "balancing test." This test said that a secular political entity, the state and/or its departments and subdivisions, had to show that its reason for intruding on religious practices had some reasonable relationship to its mission and that it had chosen the least intrusive way of doing business—otherwise state action could violate the constitutional prohibition.

It was not long after the second decision, involving an Indian in Oregon who was a member of the Native American Church and had admitted to using peyote in a ceremony, that a great coalition of American religious bodies came together to seek reversal of the Supreme Court ruling on their own behalf. In the fall of 1993 they were able to get Congress to pass the Religious Freedom Restoration Act, releasing them from the then-plenary powers of the states to do whatsoever they liked with and to religious bodies and their people.

Indians were not welcomed at the coalition meetings and every effort was made to distance the mainstream religious bodies from any involvement with Indian religions. The shortsightedness of this approach was astounding to anyone who had followed the issue. Until the Rehnquist court began its attack on Indians, there was little involvement of Indian religious cases in First Amendment law. Indeed, it would have been an embarrassment for the Supreme Court to have been forced to examine certain Indian cases in which former justices had declared that Christian ethics and values should be established as the standard by which Indians should be treated.

It should have occurred to church leaders that having once embroiled Indians in First Amendment case law, the Rehnquist court would not hesitate to do so again. Excluding Indians from the remedial legislation, therefore, was useless since the first Indian religious freedom case to reach the Supreme Court would simply revive the issue and further damage constitutional protections for all religious traditions.Thus to omit Indians from their bill and to believe they had escaped the clutches of the judiciary was extremely naive. If the Rehnquist court was determined to destroy First Amendment law in order to attack Indians, *any* statute would be vulnerable to attack—unless the Court ruled that the statute applied only to Christians and Jews—which would clearly be establishing these religions.

Thus, while the Christian churches are supine for most purposes, and in fact seem to have destroyed all respect for the traditional tenets of religion, it is still necessary to deal with Christianity because it serves as a means of organizing a large number of Americans for a wide variety of purposes. Evidence of the organizing ability of the institutional Christian churches was copious in the 1994 congressional elections. Traditional fundamentalist doctrines such as a strong military, no welfare, unlimited access to guns, and prayer in school brought out a tidal wave of voters to place the Republicans in control of the Congress again. The seeming inconsistency of being "pro-life" and for the death penalty and voting for a convicted felon for senator in Virginia in an effort to restore morality to the nation's capital escaped most voters.

If Christianity had become a powerful political force in the United States, it was also abandoning its traditional posture as a religion. In Colorado in 1994 a coalition of churches filed an amicus brief to the effect that while they were responsible for licensing clergy, and held out publicly that they and their ministers represented the Supreme Deity of the universe, they should not be held liable when their clergy violated their ministerial oaths and sexually abused their parishioners. Not even Jim Bakker or Jimmy Swaggart had the nerve to offer this argument to a court of law. Catholic bishops met later in the fall and sought out ways to reduce the abuses wrought by the clergy on their flock, but no apparent progress was made.

American Indians then have been left in the lurch by a society that craves participation in their ceremonials but which will not provide a scintilla of protection for their religions. In hearings on legislation to provide protection for traditional tribal religions, representatives of the Justice Department at one time seriously suggested that the only way they could support the legislation would be to remove the word "religion" from it. They expressed great fears that the Indian protections, and I must here stress the idea of protections although damned little was provided, would "establish" tribal religions. It was a case of legal logic run amok. To listen to representatives of the Clinton administration, one would suspect that a large war party of Sioux Indians was waiting just beyond the beltway, ready to invade the District of Columbia and force members of the executive branch to participate in the Sun Dance.

"Religion" in the original sense of the Constitution means the various Christian denominations, whose members and clergy had been active in religious persecution in Europe and who might, given some authority, repeat their tyranny in America. The First Amendment was designed to keep Christians from killing each other. It was *not* designed to keep a handful of old Indians from saying prayers in the California High Country or to prevent another handful from ingesting some peyote buttons in a remote setting in Oregon. Courts and politicians seem to miss these simple distinctions and ramble incoherently about "favoring" a religion. The truth is that practitioners of the traditional religions are not seeking converts and have no goal of gaining control of the police powers of the state to force people to say their prayers. Indeed, they flee from involvement with people outside their own group at the slightest provocation.

Common sense should demonstrate that aggressive missionary religions intent on forcing their views on the rest of society should not be established and religions that do not seek converts or involvement with political structures should not have the establishment clause raised against efforts to guarantee them a chance to worship and conduct ceremonies. The final irony of this situation is that America was once a nation where religious freedom was guaranteed. Now it is guaranteed to everyone except the original inhabitants, who have never demonstrated religious intolerance at all.

The area of government is as confused as anything we can identify, and one need only look at TV coverage of either House or Senate to see that we do not have a government at all but merely a group of fat cats who spend their time, and our money, being nasty to each other. For nearly a century it was almost standard practice to indict the attorney general of each Republican administration because he had been the bagman for the corruption practiced by that party. In recent years political appointees of both parties have had such checkered backgrounds that we have had to create a special prosecutorial function, the real fourth branch of government, for the investigation and prosecution of a good percentage of people who have served in government. Some years ago a person in Maryland ran for office on the platform that he had been to jail already and so he wanted to go to Congress and reverse the traditional pattern.

Indians suffer severely with each new administration because, in spite of campaign promises, they are always far down the list of interest groups that need attention. The story goes that when the Carter administration was handing out assignments, the subject of allocating responsibility for Indians among the White House staff came up and the group winnowing through the resumes drew a blank. Finally, one of the people alleged that his grandmother was a Cherokee, which is really just another way of saying that a person was born in the United States, and he got the job.

With affirmative action, it has been the practice to find Indians to fill the top positions of agencies that serve Indians. Administrations rarely go beyond the beltway to inquire about the qualifications of possible appointees, and consequently these jobs are often filled with people having a remote genetic relationship with an Indian tribe or with Indians who will do almost anything to ensure keeping a job. The last person to hold the commissioner of Indian Affairs position in the federal government who had any understanding of the nature of the job was a white man—Philleo Nash of Wisconsin. His successor, Robert Bennett, an Oneida and a lifetime bureaucrat, did a credible job, but most of the succeeding appointees, unfortunately all Indians, have been a singularly undistinguished lot who wanted nothing more than to ride in government limousines and be given testimonial dinners.

Appointment to Senate or House Indian subcommittees has been a fate most politicians avoid like virtue. (It would be absurd to describe an American politician avoiding sin.) In the early 1960s the Senate Indian Subcommittee had a series of reasonably competent and experienced leaders—George McGovern, Lee Metcalf, Paul Fannin. The House had James Haley, who worked hard to unravel the mess that has become federal Indian law. Both Robert Kennedy and Edward Kennedy did a great deal for Indian education, but in general the last thirty years have seen stumbling, bumbling, and simple disclaimers by members of the committees that Congress should let the Bureau of Indian Affairs do their thinking for them.

Senator Daniel Inouye of Hawaii chaired the committee in recent years until the Democrats lost control of the Senate, and his leadership was outstanding. He is very knowledgeable about Indians and has sought earnestly to bring about a general reform of federal Indian law. From a historical perspective, it is fair to say that no other senator has given so much to this cause as Senator Inouye. His chief antagonists, however, have been educated Indians working for the Department of the Interior who received their jobs thanks to affirmative action and who see their mission as one of protecting the powers of the BIA over Indians. Other than Senator Inouye, however, few senators or representatives even bother to attend committee hearings in either the Senate or the House. Consequently, much legislation is held up while last-minute deals are consummated by inattentive lawmakers.

During the Reagan administration, Indians feared a return to the old Republican desire to terminate the federal relationship and so sought some reassurance that programs would not be cut. The slogan tribal leaders used was a clever one. They wanted to maintain a "government-to-government" relationship; in other words, they wished to be treated as if they were a regular government within the American political framework. The Clinton administration, in its flashy but meaningless summit meeting with tribal officials, emphasized this slogan as the watchword of the new administration.

Unfortunately, the phrase became so popular that by the summer of 1994 Clinton appointees were using the phrase as if it were embedded in the Constitution as a clear and unambiguous part of one of the articles. Government Indians rushed from meeting to meeting chanting "government to government" in a manner reminiscent of medieval monks really getting into the mass. Rumors started to fly regarding the meaning of this policy—at least the meaning for a good number of the Indian bureaucrats. A wave of fear spread through government circles as it was suggested that there was *no* federal responsibility for Indian health, education, or welfare. The bureau Indians, who had taken advantage of federal finances to support their undergraduate and graduate education, even argued that there was no trust responsibility for individual Indians, only this nebulous and glorified "government-to-government" arrangement.

It should not have taken people long to ask why, after over two hundred years of receiving services in education, health, and other areas, it was left to the Clinton administration to ferret out the lack of authority to do so. Did more than one hundred Congresses, forty presidents, and a host of Supreme Court justices not know as much as young Clinton appointees? When cornered by inquiring Indian representatives, the articulators of these startling new theories of law solved the problem by denying they had ever voiced such thoughts. Unfortunately, most of the inquiring Indians were too polite to remind the Justice and Interior representatives of their earlier statements.

The relationship between the federal government and the Indians is complex. It has always been complex, and generations of thinkers have tried to give permanent shape and substance to it and have failed. Basically, it begins in the centuries when international law was being created to solve a number of the stormier problems created by the era of colonialism. The pope, as Vicar of Christ and therefore arguably the landlord of the planet, had disrupted international politics within months of the discovery of the New World by issuing a bull purporting to give most of the Western Hemisphere to Spain in exchange for the Spanish guarantee that the newly discovered Natives would be converted to the faith. Succeeding popes also issued bulls, and reading the perspectives that these missives contained it is not difficult to understand why they were called "bull," as they gave away the property of the Natives and enslaved the peoples of non-Christian societies all over the world.

It was not hard to disclaim any sense of national status or political standing for small tribes in the Western Hemisphere who could not mount significant military resistance to the intrusions of Europeans. The obvious civilized state of some of the Asian countries precluded the argument that they were uncivilized. Thus, doctrines of political supremacy and the corresponding responsibility to bring non-Christian nations within the circle of civilized life through education and various forms of mercantile exploitation meant that the relationship between the major colonial powers and the smaller indigenous peoples of the Western Hemisphere was never clarified.

Christians were supposed to bring non-Christian peoples into the faith, but what exactly did that mean? The Spanish argued that this divine commission justified the enslavement of Indians, contending that the rigorous religious discipline of the Church and an unbearably hard life as laborers in Spanish mines, on farms, and ranches, and as serfs for the mission, brought the Natives closer to God. This same excuse was used by southern planters in their tongue-in-cheek apologies for African slavery.

Chief Justice John Marshall tried through analogy to describe the nature of the federal-Indian relationship by saying that it resembled the relations of a guardian and a ward. Subsequent generations of American politicians adopted this language as their excuse for stealing Indian lands, arguing that the Indians were legally incompetent—an argument that did not hold much weight when the tribes ceded lands, since on those occasions the Indians were regarded by the law as highly skilled negotiators. Federal case law has fluctuated back and forth between describing Indian tribes as quasi-independent and stipulating them as wards of the government.

With the allotment of reservation lands, "trust"—the technical term for the government's responsibilities toward Indians—took on a new twist. Vowing to "break up the tribal mass," the Cleveland, Harrison, McKinley, and Theodore Roosevelt administrations forced agreements on most of the western tribes to subdivide their lands and sell the "surplus" to the government for settlement by whites. The justification for breaking up the tribal land estate was to enable individual Indians to have land of their own—thus taking advantage of the great civilizing force of private property—and to enable Indians to learn how to manage property. Generally, twenty-five-year periods of trust were placed on individual Indian lands so that people would not lose their farms before they became capable of using them.

In 1891, just as the allotment process was getting under way, Congress amended the allotment law and gave the secretary of the interior the power to manage allotments owned by minors, the elderly, and anyone who could not take care of his or her property. In effect, the generalized responsibilities of the government now became a matter of placing the property of individuals under a trust that much more closely resembled an ordinary trust arrangement in regular domestic law. Charles Curtis, when he was vice president of the United States, could not manage his own allotment because he was a Kaw Indian. Since that time, the Bureau of Indian Affairs has done very little *except* manage lands for individuals. To have young Indian lawyers then scurrying around Washington saying that there was no trust for individuals and that the proper relationship was simply "government-to-government" made no sense at all. It was, in fact, rather alarming.

The implications of restricting government services to tribal governments proved ridiculous when Interior Department representatives tried to argue for their version of the legislation. They wanted religious freedom protections restricted to federally recognized tribes—as if a tribe could have a religious experience. There was one problem with the Interior position, and that was that being recognized as an Indian was a reasonably simple thing: one need only allege it as a fact and most courts would not inquire further. Prior to the two devastating Supreme Court cases, another Indian religious freedom case had been heard by the Court in which an Abnaki Indian alleged that the issuance of a social security number to his daughter would injure her soul. As it turned out, after the initial trial he admitted that his daughter already had a social security number, but he appealed anyway. The Supreme Court, as might be expected, ruled against him in *Bowen v. Roy* (1986).[3] The Abnakis were not recognized by the federal government as an Indian tribe; however, the case was described by everyone concerned as an "Indian" case, forecasting the impossibility of restricting the future actions of a state or federal court in limiting the results of cases to federal Indians or tribes.

Toward the end of 1993, in a fit of inspiration, the Bureau of Indian Affairs took the position, wholly without merit, that the Indian tribes could be classified as "historic" and "nonhistoric" and that the nonhistoric tribes had considerably fewer sovereign powers than the historic tribes. According to the bureau definition, the multitude of tribes classified as nonhistoric included any reservation that had more than one tribe living on it, any tribe that had been removed from its original location to another place, and any tribe that had been recognized by legislation or by the process authorized for adopting constitutions by the Indian Reorganization Act. In short, these people simply threw away existing federal Indian law and were determined to begin again with their own criteria for recognizing Indians. Even the most vicious anti-Indian people in the past two decades had always followed the law, but the Clinton administration either did not know the law or was not interested in following it.

After some considerable effort, Senators Daniel Inouye of Hawaii and John McCain of Arizona put a rider on a bill that prohibited BIA personnel from changing the law by writing solicitor's opinions. This prohibition was welcome, of course, but it brought up an entirely new problem. Was it really necessary to instruct federal employees that they did not have a constitutional right to change the law to fit their needs? The Interior doctrine was something akin to Richard Nixon explaining to David Frost that if the president does something that would ordinarily be illegal, because he is the president it is not illegal. While we have had great jurists in American history—Louis Brandeis, Benjamin Cardozo, John Marshall, Hugo Black, Oliver Wendell Holmes, and Felix Frankfurter—it does appear that Lewis Carroll has carried the day.

We have three terribly complex areas in which Indians have a radically different situation than the rest of American society. The problems, although they seem insurmountable, are really quite simple if honest efforts are made to correct the misconceptions and injustices. Indeed, the people who man the barricades in science, religion, and politics have one thing in common that they do not share with the rest of the citizenry. They are responsible for creating a technical language, incomprehensible to the rest of us, whereby we cede to them our right and responsibility to think. They in turn formulate beautiful sets of lies that lull us to sleep and enable us to forget about our troubles, eventually depriving us of all rights.

If you think for a moment about the last time you heard a scientist, a minister, a politician, or a lawyer explain something, you will come to realize that we have not been told anything we did not already know. But it has been explained to us in such a convoluted manner that we are damned if we understand what they are trying to say. Many people will remember the Carter years when we had high unemployment, high inflation, and high interest rates—an impossibility under orthodox economic theory, yet no economist could explain how these three economic indicators could be high simultaneously. While no one could explain it, we nevertheless heard a symphony of big words provided in such soothing tones that we missed the dreadful inconsistency involved.

This book deals with some of the problems created for American Indians by science. We will encounter a number of amazing inconsistencies in the manner in which science describes the world we live in and the role it has chosen for American Indians to play in a largely fictional scenario describing prehistoric North America. It is not enough, however, to demonstrate the fallacies of Western science. I will offer an alternative view of North American history as seen through the eyes and memories of American Indians.

2

Science and
the Oral Tradition

EVERY HUMAN SOCIETY maintains its sense of identity with a set of stories that explain, at least to its satisfaction, how things came to be. A good many societies begin at a creation and carry forward a tenuous link of events, which they consider to be historical—which is to say actual experiences of the group that often serve as precedents for determining present and future actions. Sometimes these stories incorporate moral teachings and what we have come to call religious traditions, the actions of the higher spiritual powers or invisible forces that were important actors in the more spectacular and memorable events of their history. A good many societies speak of catastrophic events or of the movement of their people from one planet to another. Monsters and strange creatures also appear in stories and beg credibility when these tales are recited.

Of those societies that found a way to create a written record of the past, the Hebrews have been most influential, since it was the adoption of the Hebrew version of ancient events that came to be accepted, through the spread of Christianity, as the valid and incontestable explanation of how this planet came to be. Arguments about the great flood of Noah and the presence in geological strata of skeletons of animals not seen today opened the floodgates of controversy about the age of the Earth and directed the attention of Western thinkers toward the proposition that our planet might have a much different past. Eventually, the believers in biblical accuracy were put to rout by secular thinkers who substituted a seemingly infinite amount of time during which everything "evolved" in place of the shorter time scale of creation and religious history as it was represented in the Bible.

Most Americans do not pause to look back at the developments of the past two hundred years that make our society and time unique. With the triumph of Darwinian evolution as the accepted explanation of the origin of our Earth—indeed, of the whole universe—we are the first society to accept a purely mechanistic origin for ourselves and the teeming life we find on planet Earth. Science tells us that this whole panorama of life, our deepest experiences, and our most cherished ideas and emotions are really just the result of a fortunate combination of amino acids happening to coalesce billions of years ago and that our most profound experiences are simply electrical impulses derived from the logical consequence of that first accident. We thus stand alone against the cumulative memories and wisdom of all other societies when maintaining this point of view. We justify our position by accusing our ancestors and existing tribal societies of being superstitious and ignorant of the real causes of organic existence. Do we really have a basis for this belief?

Unfortunately, the discussion of the age of the Earth and the nature of past events has been conducted wholly within the confines of Western civilization. Consequently, the traditions of all other peoples have been shunted aside, since, if the Bible were shown to be mythical fairy tales, and it was the confirmed word of God, the accounts of other peoples, non-Westerners, would be even less reliable. When secular science defeated Christian fundamentalism, in its victory it was able to promulgate the belief that *all* accounts of a creation or of spectacular catastrophic events were superstitions devised by ignorant peoples to explain the processes of the world around them. The defeat of Christianity foreclosed the possibility that any other tradition that had accounts of past Earth events could join in the enterprise to explain to an increasingly global society the origins of the planet and of our race.

Since Christianity was regarded as the only true religion, all traditions were seen as folklore and myth.

Religious people did not surrender easily and, indeed, even today we have a surprising number of people who believe the literal truth of Old Testament accounts and reject or grudgingly admit the possibility of the secular scientific explanation. The scientific view of Earth history is a rather simple thing depending more on the status of secular science than on the validity of its tenets. Science simply demanded an infinite amount of time during which minute changes in both organisms and geological strata could occur. Given an infinite amount of time, and then promulgating the thesis that all change was incremental, indeed almost infinitesimal, generations of scientists were able to convince us that we "evolved" from apes, that our geological strata, the large limestone and sandstone mountains that we love so much, were the product of changes so small that they could not be detected by present-day observations. No good overview of scientific beliefs has been presented since the defeat of the fundamentalists. Consequently, the belief in infinitesimal change has received little critical examination and contains numerous errors of both logic and interpretation.

Inconsistencies abound, but we are so brainwashed by science that we do not even ask the relatively simple questions about ordinary things. We have shifting continents attached to sliding gigantic "plates" and we also have continents rising and falling to allow for the deposition of limestone and marine sediments. Exactly how both processes can occur at once is not clear, and it is only our trained belief that an infinite amount of time has passed that allows both processes to be held as literal truths. Species both evolve and stop evolving during this time period. Read any book on evolution and you will discover a surprising number of modern species that stopped evolving millions of years ago. In addition, even the most sophisticated of modern scientists, in explaining the fossil remains, finds that species in the rocks are distant relatives to each other, not direct lineages.

Our modern scientists, led by Stephen Jay Gould, have devised a new explanation of evolution that reverses Darwin's original ideas completely. Where secular science once pointed at fossil formations as evidence that evolutionary changes occurred over time, Gould and his friends point out the *absence* of transitional fossils as evidence that evolution occurred, but in rapid spurts. Apparently somewhere, and at a time unknown, when species were ready to evolve they went offstage, made their changes, and then rushed back into the geologic strata to leave evidence of their existence.

Thus millions of years are believed to have passed between species of similar body form although no real evidence of the passing of years is available. The Grand Canyon of Arizona, touted as one of the locations where we can see this infinite time pass, lacks two major geological formations—the Ordovician and Silurian—and we have no explanation for these profoundly long missing periods of geological time. We are told that dinosaurs lived 65 million years ago, and yet from Roy Chapman Andrews forward our scientists continue to find the dinosaur skeletons on top of the ground or very near the surface. So 65 million years of sediment apparently was insufficient to bury these monsters from our sight.

Any group that wishes to be regarded as the authority in a human society must not simply banish or discredit the views of their rivals, they must become the sole source of truth for that society and defend their status and the power to interpret against all comers by providing the best explanation of the data. As priests and politicians have discovered, it is even permissible to tell lies in order to maintain status, since the most fatal counterattack against entrenched authority will not be directed against their facts but against their status. As Americans, we have been trained to believe that science is infallible in the sense that, while science does not know everything, its processes of investigation and experimentation are the best available so that, given time and resources, the truth will eventually be discovered. This belief has degenerated into a strange form of religious belief because the technology that science provides us, best exemplified in the "instant replay" in sports, encourages us to cede all critical faculties to science in exchange for creature comforts.

Like any other group of priests and politicians, however, scientists lie and fudge their conclusions as much as the most distrusted professions in our society—lawyers and car dealers. In recent years we have seen several instances of false or doctored research reports, one a very serious manipulation of cancer research on women's breasts. It appears that, earlier, Charles Darwin liberally borrowed the ideas of Alfred Wallace to become the father of evolution. Johannes Kepler fudged his math in order to get the scientists of his day to take his theories seriously, and it appears that Gregor Mendel's plants did not always follow the genetic doctrines that he later articulated. Louis Pasteur, a giant of science, apparently "lied about his research, stole ideas from a competitor and was deceitful" in some fundamental ways, according to a recently published book, *The Private Science of Louis Pasteur,* by Gerald L. Geison. Lawrence K. Altman, in a review in the *The New York Times Book Review,* observed that disclosures made by the book "... are revealing that science is not as objective, neat and scrupulously honest as it is portrayed."[1]

Perhaps the epitome of scientific fraud was the work of Sir Cyril Burt on twins. Fearful of peer criticism of his work, Burt simply performed the peer-review process by himself, writing glowing reviews of his work using pseudonyms. This deceit, and the manipulation of statistical data in the studies, was eventually exposed, but in the interim most fellow scientists simply followed the party line and supported him because he was one of the old-boy network. And severe questions arise whether or not Teilhard de Chardin was a part of the Piltdown fraud that substantially affected articulation of theories of human evolution for half of this century.

We are taught to visualize the scientist as a cheerful fellow clad in a white smock, working in a spotless lab, and asking the insightful questions that will eventually reach us at K-Mart in the form of improved vitamins, new kinds of audiotapes, and labor-saving

devices. On reaching the end of his experiment, which has featured a set of daring questions that he is forcing Mother Nature to
surrender, our scientist publishes his results. His peers give serious critical attention to his theory and check his lab results
and interpretations, and science moves another step forward into
the unknown.

Eventually, we are told, the results of this research, combined
with many other reports, are digested by intellects of the highest
order and the paradigm of scientific explanation moves steadily
forward, reducing the number of secrets that Mother Nature has
left. Finally, popular science writers—Stephen Jay Gould, Carl
Sagan, Jared Diamond, Robert Ardrey, and Jacob Bronowski—and
others take this mass of technical scientific wisdom and distill it
for us poor ignorant lay people so we can understand in general
terms the great wisdom that science has created.

The actual situation is much different. Academics, and they include everyone we think of as scientists except people who work
in commercial labs, are incredibly timid people. Many of them are
intent primarily on maintaining their status within their university and profession and consequently they resemble nothing so
much as cocker spaniels who are eager to please their masters,
the masters in this case being the vaguely defined academic profession. Scholars, and again I include scientists, are generally spe-
cialists in their field and are often wholly ignorant of developments
outside their field. Thus, a person can become an international
expert on butterflies and not know a single thing about frogs other
than that they are disappearing—a fact more often picked up in
the Sunday newspaper science section than from reading a scientific journal. Scientists and scholars are notoriously obedient to
the consensus opinions of their profession, which usually means
they pay homage to the opinions of scholars and scientists who
occupy the prestige chairs at Ivy League and large research universities or even dead personalities of the past.

Scientists do work hard in maintaining themselves within their
niche in their respective disciplines. This task is accomplished by

publishing articles in the journals of their profession. A glance at the index of any journal will reveal that the articles are written for the express purpose of generating mystique and appear to be carefully edited to eliminate any possibility of a clear thought. Editors of journals and editorial boards are notoriously conservative and reject anything that would resemble a breath of fresh air.

Any idea that appears to challenge orthodoxy and is published is usually accompanied by copious responses from the names in the profession who are given an opportunity to quash any heretical conclusions that the article might suggest. Many subjects, no matter how interesting, are simply prohibited because they call into question long-standing beliefs. Prestigious personalities can determine what is published and what is not. Journals do not reflect science or human knowledge; they represent the subjects that are not prohibited in polite discussion by a few established personalities in the larger intellectual world.

We often read newspaper accounts of new scientific theories. Too often we have been trained to believe that the new discoveries are proven fact rather than speculative supposition within a field that is already dominated by orthodox doctrines. Quite frequently the newspaper accounts will contain the phrase "most scientists agree," implying to the lay person that hundreds of scientists have sincerely and prayerfully considered the issue, reached a consensus, and believe that the theory is reliable.

Nothing could be further from reality. In all probability a handful of people have read or heard of the article and, because it is written by a "responsible scholar," have feared to criticize it. But who is the responsible scholar responsible to? Not to the public, not to science, or history, or anthropology, but to the small group of similarly situated people who will make recommendations on behalf of his or her scholarship, award the prizes which each discipline holds dear, and write letters advocating his or her advancement. Unless a "scientist" is speaking specifically about his or her field, the chances are very great that he or she does not know any more about the subject than your average well-read layperson.

Since it is possible for a prestigious personality to dominate a field populated with fearful little people trying to protect their status, some areas of "science" have not progressed in decades and some scientific doctrines actually have no roots except their traditional place in the intellectual structure of the discipline. For more than a century scientists have labeled unknown animal behavior as "instinct," which simply indicated that they did not know the processes of response. And instinct was passed off as a responsible scientific answer to an important question. "Evolution" is used to cover a multitude of academic sins.

Samuel Eliot Morison was a singularly devoted worshipper of Columbus, and while he was alive it was virtually impossible to discuss pre-Columbian expeditions to the Western Hemisphere in any academic setting. It is still anathema to give the topic serious consideration. Aleš Hrdlička, longtime anthropologist at the Smithsonian Institution, was a zealous foe of early dates for the populating of North America, and even today most anthropologists and archaeologists immediately run to their computers to discredit any digs that would suggest a date earlier than 12,000 B.C.–50,000 B.C. The recent findings at Monte Verde in Chile were actually the reluctant admission by the dinosaurs of archaeology/anthropology who grudgingly agreed to add 1,300 years to the acceptable date of human occupation of the Western Hemisphere—not a notable "advance" in scientific thinking considering the inaccuracies of c-14 dating techniques.

I came to academia believing the traditional picture of the scholar after nearly two decades of political activism. Joining the University of Arizona Department of Political Science, I was very interested in learning more about political theory, since I had heard good arguments made during the civil rights movement on the implications of social contract theory. For over a year I patiently went to lunch with my colleagues hoping to engage in heated analyses of the central tenets of our profession. I learned mostly about the politics of getting an edge with the administration for more benefits, a great deal about the adjustments being made in the athletic program, and a bit about the campus affairs of the more active administrators and professors. I do not recall any conversation in which anything of substance dealing with our discipline was forthcoming.

Arriving at the University of Colorado, I was stunned to hear from my students that some of my history colleagues were beginning their courses on American history with a mindless recitation of the Bering Strait theory of the peopling of the Western Hemisphere. Basically, they were simply repeating scholarly folklore, since there is, to my knowledge, no good source which articulates the theory in any reasonable format. Indeed, this "theory" has been around so long that people no longer feel they have to explain or defend it—they can merely refer to it. I will deal with this issue in another chapter. It is important here simply to note that the immense knowledge and factual proof of many scientific theories does not exist. Many theories and facts recited by scholars and scientists today are merely academic folklore which professors heard in their undergraduate days and have not examined at all.

Some forty-five years ago, Immanuel Velikovsky published his classic work *Worlds in Collision*, in which he suggested that the Earth had been subjected to several catastrophes of an extraterrestrial nature that had involved Mars and Venus. He based these ideas on the Old Testament memories of the Hebrews and added an immense number of footnotes referring to the memories of other peoples. A significant number of his suggestions regarding the nature of our solar system and the geological features of the planets have been proved accurate in the decades since he wrote. By and large, however, scientists rushed to attack his books and threatened a boycott of Macmillan, his publisher, which made it necessary to move the book to Doubleday to keep it in print.

A sufficient literature has evolved since then to argue the case for Velikovsky, but I would like to illustrate the scientific response in just one instance because it demonstrates the tenacity with which the academic community holds on to its beliefs. Velikovsky said that Venus at one time had been a comet and had disrupted

Earth. On the basis of this identification he suggested that the surface temperature of Venus would be something approaching incandescence. Orthodox science at the time believed that Venus had a surface temperature of 25°C.

With the space probes able to gather considerably more information on Venus, the surface temperature when measured was estimated at around 800°C—a substantially radical difference. When Velikovsky pointed out the difference in measurement, his critics replied that heat was a "relative" term. Today, this high temperature is explained by an ad hoc "greenhouse" theory which suggests that clouds can raise the temperatures of planets to incredibly high measurements by a natural process.

Much of the documentation used by Velikovsky relied on the recorded beliefs of non-Western peoples in every part of the globe. Often his technique was to seek traditions that would involve some discernible physical change in a local environment that might be anticipated if a much larger and more violent event affected the whole planet. An example of this practice was to take the Long Day of Joshua and look at the other side of the Earth and find a tradition in Central America in which the night was extended for a prolonged duration. He found corresponding evidence for several of the events of his interplanetary collisions in the traditions of tribal peoples.

Some evidence, and I will review these traditions in a later chapter, did not offer much support for the cometary disruption as he conceived it. Nevertheless, by incorporating these non-Western traditions into a theory in which events having a planetary scope could be suggested and evidence for such an event could be examined, Velikovsky offered a scenario in which a truly planetary history could be constructed.

Orthodox science has done just the reverse. It accepts non-Western traditions to the degree to which they help to bolster the existing and approved orthodox doctrines. The vast majority of the time, the non-Western interpretations of Earth history and the history of human beings are rejected as Stone Age remnants of human societies that could not invent or accept the mechanistic and later industrial interpretation of the natural world. The evolutionary framework we presently have does not represent human experiences of the past and present, but simply Western doctrinal arrangements of selected bits of evidence of those experiences.

Respect for non-Western traditions is exceedingly difficult to achieve. Not only did secular scientists rout the Christian fundamentalists, they placed themselves in the posture of knowing more, on the basis of their own very short-term investigations, than the collective remembrances of the rest of humankind. Social science, in particular anthropology, preserved information about the remnants of tribal cultures around the world, most particularly in North America, but it also promulgated the idea that these tribal cultures were of Stone Age achievement and represented primitive superstitions which could not be believed.

It was with a certain degree of satisfaction, therefore, that I watched the comet crash into Jupiter during the summer of 1994, since it has been orthodox science for over a century that our solar system is immune to radical disruption by outside cosmic bodies—one of the charges made against Velikovsky. The new chaos theory, now one of the popular ways of examining phenomena, suggests that constant uniformity is probably not a characteristic of any system in this universe. The event and the new theory lend considerable support for a re-examination of the insights and knowledge of tribal peoples when trying to understand the nature of our world.

Some efforts have already been made in a number of fields to investigate the knowledge of tribal peoples and incorporate it into modern scientific explanations. Thor Heyerdahl was one of the first people to show, by repeating the event, that ancient peoples could well have traveled by sea to various parts of the globe. I think partially as a result of his voyages a small group of anthropologists have now allowed that Indians, instead of marching four abreast over the mythical Bering land bridge, might have come by boat on a bay and inlet basis from the Asian continent to North America.

Recognizing that Indians may have been capable of building boats seems a minor step forward until we remember that for almost two centuries scientific doctrine *required* that Indians come by land because they were incapable of building rafts. Polynesian voyages of considerable distances have now been duplicated, giving credence to the idea that Hawaiian tales of sea voyages were not superstitious ways of discussing ocean currents. Critical in this respect is the fact that Hawaiians would not be believed until a white man had duplicated the feat.

In methodological terms there is a major problem in bringing non-Western traditions within the scope of serious scientific perspective, and that is the inherent racism in academia and in scientific circles. Some of the racism is doctrinaire and unforgiving—for instance, the belief that, for a person and/or community possessing any knowledge that is not white/Western in origin, the data is unreliable. A corollary of this belief is that non-Western peoples tend to be excitable, are subjective and not objective, and consequently are unreliable observers.

Other attitudes encompass the idea that the non-Western knowledge, while interesting, is a lucky correspondence between what science has "proved" and what these people discovered by chance. Even with tribal peoples now entering academic fields, there is bias, and most academics deeply believe that an Indian, or any other non-Western person, cannot be an accurate observer of his or her own traditions because that individual is personally involved. It follows, to listen to the apologists for many university departments, that an urban, educated white person, who admittedly has a deep personal interest in a non-Western community but who does not speak the language, has never lived in the community, and visits the people only occasionally during the summer, has a better understanding of the culture, economics, and politics of the group than do the people themselves. When this attitude is seen in religious studies it is appalling: white scholars truly believe that they know more about tribal religions than the people who actually do the ceremonies.

The bottom line about the information possessed by non-Western peoples is that the information becomes valid only when offered by a white scholar recognized by the academic establishment; in effect, the color of the skin guarantees scientific objectivity. Thus ethnic scholars are not encouraged to do research in their own communities—studies done by whites are preferred. Many scholars with ethnic backgrounds are even denied tenure because they are ethnic and their studies and publications relate to that background. Particularly in the arts and social sciences, supposed bastions of liberalism, minority scholars are simply run out of the professions unless they are totally submissive to prevailing doctrines of the discipline and their writings do not clash with established authority.

We come then to examine the content of "science" and the "oral tradition," which is to say the traditions of non-Western peoples. Arnold Toynbee in *A Study of History* criticized his discipline for its parochial perspective. He wrote that it was "as though a geographer were to produce a book entitled 'World Geography' which proved on inspection to be all about the Mediterranean Basin and Europe."[2] By analogy, "science" is pretty much the same. It is that collection of beliefs—some with considerable evidence, some lacking any proof at all—which reflects data gathered by a small group of people over the past five hundred years with the simple belief that phenomena have been objectively observed and properly described because they have sworn themselves to sincerity. Unfortunately the assumptions and presuppositions which these people bring to the interpretation of phenomena are regarded as "normal"—as the way that people validly view the world.

Anomalies, facts that cannot or do not fit into the complete edifice, are simply ignored, their champions discredited. Validity and verification in science primarily consist of a willing conspiracy among scientists not to challenge the authorities in the field and to take the sincerity of colleagues as insight. Consequently, there are literally millions of observed facts which simply do not appear in scientific writing because they would tend to raise doubts about the prevailing paradigm.

The non-Western, tribal equivalent of science is the oral tradition, the teachings that have been passed down from one generation to the next over uncounted centuries. The oral tradition is a loosely held collection of anecdotal material that, taken together, explains the nature of the physical world as people have experienced it and the important events of their historical journey. The Old Testament was once oral tradition until it was written down. Sagas and Eddas form part of the European oral tradition. Some romance has attached to Indian oral traditions in recent times due to the interest in spirituality, and consequently some people have come to believe that oral traditions refer *only* to religious matters. This description is not true. The bulk of American Indian traditions probably deal with commonsense ordinary topics such as plants, animals, weather, and past events that are not particularly of a religious nature.

Until Indian tribes, and by extension other tribal peoples, were submerged by the invasion of Western colonizing peoples, the oral tradition represented not simply information on ancient events but precise knowledge of birds, animals, plants, geologic features, and religious experiences of a particular group of people. Sometimes the visions of different tribes would match and describe a particular event, experience, or condition and sometimes they would not.

Tribal elders did not worry if their version of creation was entirely different from the scenario held by a neighboring tribe. People believed that each tribe had its own special relationship with the superior spiritual forces that governed the universe. The task of each tribe was to remain true to its special calling without worrying about what others were doing. Tribal knowledge was not fragmented data arranged according to rational speculation. It was simply the distilled memory of the People describing the events they had experienced and the lands they had lived in. Black Elk, talking to John Neihardt, explained the methodology well: "This they tell, and whether it happened so or not, I do not know; but if you think about it, you can see that it is true."[3] The oral tradition, people felt, was serious; it was knowledge, and even the most unlikely aspects might be understood as true.

Fragmentation of human knowledge by science means that most explanations must be constructed on an ad hoc basis with the hope that use of the scientific method will guarantee that all bits of data are ultimately related. Unfortunately, the day of the philosopher in Western society has passed and no single group today serves the function of surveying the totality of knowledge and trying to bring it into a coherent and simple explanation which can be made available to the rest of society. Earlier in this century, scientific philosophers such as Alfred North Whitehead, P. W. Bridgman, R. G. Collingwood, Hans Margeneau, and Werner Heisenberg attempted to frame coherent explanations of the whole, relying primarily on their knowledge of physics, mathematics, and astronomy to describe the rest of the data that other disciplines of science had gathered.

In the old days, elders performed a similar function and recited the oral traditions of the tribe during the wintertime and as a regular part of camp or village life. Religious ceremonials generally involved the recitation of the origin and migration stories, and most of the accumulated wisdom of the tribe was familiar to everyone. Special knowledge regarding other forms of life, if revealed in visions or dreams, was made available to the larger community on a "need-to-know" basis, since it was generally regarded as personal knowledge.

Storytelling was a precise art because of the nature of Indian languages. Some tribal languages had as many as twenty words to describe rain, snow, wind, and other natural elements; languages had precise words to describe the various states of human emotion, the intensity of human physical efforts, and the serenity of the land itself. If the stories began "Once upon a time ..." they quickly gave the listener a completely accurate rendering of a specific experience which Western languages could not possibly duplicate. In this context, everyone understood the philosophical overview, and ad hoc explanations were treated as facts that must be understood but whose time for understanding had not yet come.

In some of the larger Indian nations, elders functioned pretty much as scientists do today. That is, no one person could remember all the information about the trivial past, the religious revelations, and the complex knowledge of the physical world. Consequently, people specialized in certain kinds of knowledge. Specialization occurred most frequently at vision quests or puberty ceremonies when young people sought help and guidance from birds, animals, and spirits. Often their careers would be shown to them and special information, roots, symbols, and powers given. This information would usually be shared with the spiritual leaders who had supervised the ceremony, but sometimes the person was told to bring a certain medicine, dance, or a bit of information to the rest of the community. The difference between non-Western and Western knowledge is that the knowledge is personal for non-Western peoples and impersonal for the Western scientist. Americans believe that anyone can use knowledge; for American Indians, only those people given the knowledge by other entities can use it properly.

Education, in the American system, is a function of class and economics, and with some rare exceptions the scientific-academic community is self-perpetuating. Middle- and upper-class peoples have a significantly better chance to become scientists than do average citizens simply because they can afford to continue in school until they receive a Ph.D. While graduate school education does provide further training in one's chosen field, its primary purpose is to ensure that people wishing to become scholars and scientists are rendered socially acceptable to people already entrenched in the respective professions.

Originally, graduate theses were supposed to be creative and original scholarly work which advanced the knowledge of the world in some significant way. In recent decades this requirement has lapsed completely, and in order to receive an advanced degree today a student need only demonstrate to the committee that he or she will not embarrass the sponsoring professors by being out-spoken or taking radical positions that would disrupt the discipline. Rarely do M.A. theses or Ph.D. dissertations contribute anything to our knowledge.

With the Plains Indian tribes, and I suspect with the vast majority of the Indian groups, the most revered person was the scout. On his knowledge and powers of observation the rest of the community vested their survival. His task was to search out herds of game animals, report the presence of enemies, analyze the weather, and be aware of the slightest change in the environment. If he was wrong, or even slightly inaccurate, the community might perish or decide on a course of action that would have detrimental effects. People sometimes decided against the course of action recommended by a scout, but they never doubted his veracity. Lying by a scout was a dreadful act punished by death or banishment.

A remarkably high percentage of scouts also became the great storytellers and were repositories of the oral tradition. They might vary some of the descriptions of events to entertain their audience, but these editing devices were recognized by everyone, since all the stories were known in their basic outline. Sometimes, in the storytelling, people vied for the chance to introduce puns and humorous variations on words which would transform the story into a multileveled account. Becoming a respected articulator of the knowledge of the tribe was not a status dependent upon economic or even military prowess. Indeed, like modern fishermen measuring fish they once caught, people tended to look suspiciously at the versions of experiences told by individuals whose accomplishments were not in the field of observation. Some tribes prohibited a person from lauding his own accomplishments for fear of exaggeration, requiring friends and relatives to describe exploits.

Comparing the two ways of gaining a position of authority in society, then, the oral tradition clearly had many more guarantees that its knowledge would not become the subject of personal bias.

The possessor of the oral traditions had nothing that would encourage him or her to change the meaning or emphasis of the information except, as already noted, the desire to entertain. People had no vested interest in wealth or prestige by becoming knowledgeable and, lacking any concept of tenure, storytellers maintained their status only to the degree that they represented information and wisdom. Any suspicion that they didn't know a subject would eliminate them as a serious and reliable source of knowledge.

Within the scientific establishment, on the other hand, immense rewards are made available to the individual who stands out among his or her colleagues. Consequently, in today's academic setting, with the impact of the television personality cult, advocating popular theories or making a theory popular is a requirement of academic success, regardless of the truth of the situation. In some instances, the more bizarre and outlandish the theory, the more useful it is in bringing economic rewards to its creator. Sensationalism often substitutes for truth, and that is one reason why we have so many popular psychologists and sociologists.

The major difference between American Indian views of the physical world and Western science lies in the premise accepted by Indians and rejected by scientists: the world in which we live is alive. Many scientists believe this idea to be primitive superstition and, consequently, scientific explanation rejects any nuance of interpretation that would credit the existence of any activities of the natural world as having partial intelligence or sentience present. American Indians look at events to determine the spiritual activity supporting or undergirding them. Science insists, albeit at a great price in understanding, that the observer be as detached as possible from the event he or she is observing. Indians know that human beings must participate in events, not isolate themselves from occurrences in the physical world. Indians thus obtain information from birds, animals, rivers, and mountains, which is inaccessible to modern science.

Again, however, there are certain kinds of correspondences between the Indian way and modern scientific techniques. We know from meteorology that seeding clouds with certain chemicals can bring rain. This method of dealing with natural forces is wholly mechanical and can be described as the power to force nature to do our bidding. Indians performed the same function by conducting ceremonies and asking the spirits for rain. Science is severely limited, however, since it cannot affect winds, clouds, and storms except by certain kinds of alterations. Luther Standing Bear recounts an instance in which a Sioux medicine man drastically changed the weather with his powers because he had become a friend to the forces that stood behind the meteorological phenomena:

> Some of my band, the Oglalas, went to visit the Brule band and by way of entertainment preparations were made for a dance and feast. The day was bright and beautiful, and everyone was dressed in feathers and painted buckskin. But a storm came up suddenly, threatening to disrupt the gathering, so of course there was much unhappiness as the wind began to blow harder and rain began to fall. Last Horse walked into his tipi and disrobed, coming out wearing only a breechclout and moccasins. His hair streamed down his back and in his hand he carried a rattle. Walking slowly to the center of the village he raised his face to the sky and sang his Thunder songs, which commanded the clouds to part. Slowly but surely, under the magic of the song, the clouds parted and the sky was clear once more.[4]

Acting in concert with friendly thunder and storm spirits is rather commonplace in many Indian tribes and demonstrates the more comprehensive scope of the oral tradition in comparison to both scientific knowledge and powers.

Indians came to understand that all things were related, and while many tribes understood this knowledge in terms of religious rituals, it was also a methodology/guideline which instructed them in making their observations of the behavior of other forms of life. Attuned to their environment, Indians could find food, locate trails, protect themselves from inclement weather, and anticipate coming events by their understanding of how entities related to each other. This knowledge is not unique to American Indians. It would be available to anyone who lived primarily in the natural world, was reasonably observant, and gave other forms of life respect for intelligence and the power of thought.

Western science also has the idea of relativity, but the concept was initially applied only in theoretical physics to explain the relationship of space, time, and matter. Gradually, scientists have moved from philosophical physics to apply the concept of relatedness to biological phenomena and environments. Now many scientists believe that all things are related, and some articles, primarily coming from people in physics, now state flatly that all things really are related. The proposition, however, still seems to be an intellectual concept that lacks the sense of emotional involvement. If scientists *really* believed in the unity and interrelatedness of all things, their emphasis would shift dramatically and they would forswear using animals for lab research, change their conception of agronomy entirely, do considerably different studies of water and landscapes, and begin to deal seriously with the by-products of their experiments. Hopefully that day is coming.

When the sciences became divided, our knowledge of the world became badly fragmented. Scientists, in creating narrow classifications of disciplines, developed more precise focus and were able to articulate the substance of the discipline and its goals. They were also able to simply discard phenomena and data which did not fit into their specialist subject area. Rejected data were called anomalies, and no single discipline assumed responsibility for including anomalies in any of the smaller disciplinary paradigms. Thus there are literally millions of irrefutable facts which science simply dismisses, even though they go to make up entities and events which composed our world.

Some areas of interdisciplinary work cause great controversy because of previous excesses by the discipline which originally had the interpretive franchise. A great many of the ancient Indian ruins in the United States were once classified as religious sites by anthropologists who never did know what they were but imagined many early cultures to be dictatorial theocracies and therefore supposed that people spent their lives building temples. Many of these same ruins are now interpreted by archaeoastronomers as primitive but sophisticated computers which can scan the horizon if properly used, and they are seen as providing proof of a complicated Indian star knowledge. At the present time we really don't know what some of these ruins represented. It is obvious that the two different interpretations simply reflect fads within Western scientific disciplines, and religious interpretations have faded for the moment. But even labeling a site as astronomical is an improvement, since it partially sidesteps the old stereotype of Indians being primitive and ignorant savages.

At the American Association for the Advancement of Science annual meeting in Chicago in 1992, there was a panel presentation of a new field called "zoopharmacognosy," which is a term describing the use of medicinal plants by animals. The panel got a laudatory review in a *Newsweek* article, which described fearless scientists spying on sick animals and observing them using certain plants to cure themselves.[5] A Duke University primatologist was quoted as saying, "If these work for primates, then they are potential treatments for humans," this insight apparently being a startling departure from ordinary scientific logic. The article quoted Harvard ethnobotanist Shawn Sigstedt suggesting that bears may have taught the Navajos to use a species of the *Ligusticum* plant, just as they had claimed!

For Western peoples, the announcement of zoopharmacognosy may be an exciting breakthrough on the frontiers of science, but getting information from birds and animals regarding plants is an absurdly self-evident proposition for American Indians. It gives substance to the idea that all things are related, and it is the basis for many tribal traditions regarding medicinal uses of plants. The excitement illustrates a point made above: Why didn't people take Indians seriously when we said that animals and birds give us information on medicinal plants? Why is such knowledge only valid and valuable when white scientists document and articulate it?

About twenty years ago a very popular book summarized some of the information being retrieved about plants—*The Secret Life of Plants* by Peter Tompkins and Christopher Bird. The book covered the various experiments done by scientists to measure what appeared to be the emotional life of plants, including the famous Backster effect in which plants' emotions were registered on a meter. A casual reference in the book to Indians using music to ensure greater plant growth was followed almost immediately by reference to a white man in Wauwatosa, Wisconsin, a florist who piped music into his greenhouses to make plants grow better.

No real discussion was ever presented regarding American Indian knowledge of plant life, even though it is well known that Corn Dances are one of the chief religious ceremonies of the southwestern Indians. In the schizophrenia that we know as America, Indians using songs and dances to improve crops is not significant, but a florist piping music into a greenhouse is astounding and illustrates a hidden principle of the universe.

We should not be too critical if some scientists are lurking in the bushes trying to discover bears eating unusual plants or if people talk to their plants, pray over them, or serenade them with music. These things do help expand the frontiers of science away from the stuffy and sterile materialistic perspective and begin to open up new ways of approaching nature. What is it, however, that blocks any possibility of dialogue between Western science and the tribal peoples who know these things, and more, as a matter of course?

Two things need to be done, in my opinion, before there can be any exchange of views between American Indians and Western science. First, corrective measures must be taken to eliminate scientific misconceptions about Indians, their culture, and their past. Second, there needs to be a way that Indian traditions can contribute to the understanding of scientific beliefs at enough specific points so that the Indian traditions will be taken seriously as valid bodies of knowledge. Both changes involve a fundamental struggle over the question of authority, since even when Indian ideas are demonstrated to be correct there is the racist propensity to argue that the Indian understanding was just an ad hoc lucky guess—which is perilously close to what now passes for scientific knowledge.

In the next several chapters I will examine some of the most ridiculous of scientific beliefs and show that they have little, if any, basis in reason or fact for being accepted. I will then look at an emerging field—geomythology—and raise questions about the validity of traditional Indian knowledge when compared with Western explanations.

3

Evolutionary
Prejudice

AN AREA OF GREAT DEBATE and considerable friction concerns
the origin of American Indians. Originally, Europeans believed that
all humans were created by the Hebrew God. Discovering previously unknown people, therefore, produced a rush to the Old Testament to discover whether or not there was some way of
identifying the new people. The options were few. Presuming that
all of humankind was represented in Noah's ark, one could trace a
path from Mount Ararat to eastern Siberia and, since the Bering
Strait was geographically adjacent to North America, to the Western Hemisphere. Although gripped in forbidding ice, snow, and
freezing temperatures much of the year, it seemed natural to posit
a migration across this quasi-isthmus. The same sequence could
be established if the scholar assumed that the Indians were the
"Ten Lost Tribes" of Israel, further assuming that instead of returning from Babylonian captivity they moved east instead of west to
help rebuild the temple.

Indian traditions also spoke of a great flood and featured tribal
ancestral heroes who built rafts and boats to escape the disaster.
Sadly, the Indian flood stories were taken as evidence of the truth
of the Bible rather than as independent evidence of a planetary
flood. It was simply assumed that Indians originated shortly after
Noah's flood and over the years got their stories garbled. And creation stories of the original state of the planet as a place covered
with water were also seen as evidence of Noah's flood, the roles of

birds and other animals in the creation of dry land being conveniently discarded so the story would match Old Testament standards. Linking Indians to the descendants of Noah meant bringing them to the Western Hemisphere via land (strangely, they forgot how to build boats and rafts), trooping across the Bering Strait.

A note of passing interest in this respect concerns Chief Joseph. After his surrender in 1877, he gave a pendant to General Miles, and this object eventually found its way to West Point. A few years ago it was examined and turned out to be a Mesopotamian tablet recording the sale of livestock, a disturbing anomaly and an undeniable fact that should have been grasped at once by Christian fundamentalists and Mormons. How this tablet got into Joseph's family and became an heirloom is a matter of some speculation, telling us that our view of Western Hemisphere prehistory is not as complete as we might think.

In the 1850s, the great debate about origins exploded with the publication of books on the theory of evolution propounded by Charles Darwin and Alfred Wallace. The major casualty in this contest was the literal interpretation of the Old Testament and, more important, the banishment of the Creator who had a sophisticated plan in favor of a benign and clever "Mother Nature" who threw dice with organic genetics and always came up a winner. Creation and a limited Earth history had already been pretty well mauled by the new science of geology, which pointed out the complexity of the fossil record and all but indicated that organic life had evolved from simple to complex. About all creationists could do was to say that all fossils were evidence of the great flood, but it was clear that there must have been a whole series of floods to create the complexity found in fossil beds.

Evolutionists made the inevitable linkage between primates and our species based primarily on the similarity of body form. Western Europeans then freely admitted their close connection with the big apes, a concession which has not yet been heard from tribal peoples and some Christian fundamentalists. Once the man-ape sequence was established, scientists then believed that a series of missing links and "hopeful monsters" had once existed, arguing that primates had eventually evolved into educated middle-class Western capitalists. It was necessary, indeed imperative, to arrange the various human societies on an extended incline in which tribal people with a crude mechanical technology illustrated the early kinds of human societies and ancient Near Eastern peoples became the predecessors of the modern industrial state, moderated eventually by the innate gentility of the Anglo-Saxon genes.

If we dig beneath the masses of evidence that Western society and its apologists throughout history have usually cited as proof of its superiority, the basic argument is that the West has been able to create more sophisticated ways to use artificial energy to perform tasks, thereby making life more enjoyable for the elite who have controlled the various political and economic institutions of the West. But this proof is not as overwhelming as its advocates like to pretend. We cannot today either *duplicate* or *explain* the means by which the ancients cut and moved large stones in the Middle East, in Central America, or on Easter Island.

We are at a loss to explain the very sophisticated astronomical knowledge of many societies. Scientific writers usually pretend that the ancient peoples were highly superstitious and that, after having created astrology, they eventually moved into a secular and objective astronomy, forgetting that at that stage of development it would have been considerably more difficult to have created an astrological horoscope than a simple map of our solar system. Since we must assume that ancient people used the naked eye to determine the planets, comets, star formations (including the color of the stars and the precession of the equinox), the ancients must have been incredibly good observers of the heavens or had access to information whose channels no longer function for us.

Tribal peoples were placed at the very bottom of the imaginary cultural evolutionary scale, and this status had two edges and cut in several directions at once. If the tribal peoples actually represented Western origins at a much earlier time, it was exceedingly valuable that they be studied intensely for clues about the nature and origin of human society. Consequently it was an injury to science and human knowledge to allow the military to simply exterminate them. But if tribal peoples represented an earlier stage of human evolution, everything they said, believed, or practiced must necessarily reflect a stage of superstition from which western Europeans had emerged. Therefore, their traditions were simply fairy tales made deliberately to explain a cosmos which they feared; their technology was a protoversion of plows, reapers, combines, and food-processing plants which we see in modern industrial society. Their memories of past experience were clouded with superstition, and their religious beliefs were the lowest stage of animism, which produced a life of fear of spirits.

Western civilization, by the time it reached the shores of this hemisphere, had pretty much institutionalized its beliefs and experiences. That is to say, problem solving was already an institutional function: people purchased food grown by others, settled their conflicts in courts and legislatures and not by informal mutually agreed-upon solutions, and waged extended and terrible wars instead of mere battles over the right to occupy lands for hunting and fishing purposes. Beliefs about the world had been processed into philosophical systems that articulated rational principles rather than anecdotal experiences, and religion had been reduced to creeds, dogmas, and doctrines.

The first go-round of real inquiry into the nature of tribal societies *assumed* that all human societies had developed a concept of property, a formal governing institution, a crude but effective system of economics designed to produce surplus wealth, and sets of formal laws, usually focused on the male, which governed domestic relations. It was further believed that all human societies began as animists—people who saw spirits everywhere—and had gradually evolved through polytheism and human sacrifices into monotheisms, which produced wonderful ethical codes that expressed in the abstract the kinds of beliefs and behavior necessary to produce a civilized society.

Needless to say, the nineteenth century saw the beginning of anthropology on a grand scale. Western nations were grabbing large parts of the globe, intruding on peoples living in remote locations on the planet who asked only to be left alone, and extending the reach of democracy and capitalism to embrace everyone. The methodology of everyone who looked at or considered the existence of tribal peoples was to find happy coincidences between tribal beliefs and practices and the way that Western peoples did business. It goes without saying that the coincidence in beliefs and practices only served to entrench the belief that all peoples *began as primitives* and *inevitably* moved toward Western forms of organization, which in turn were guaranteed by Western religion and philosophy, which had themselves survived thousands of years of criticism and refinement. By wrapping cultural evolution so tightly, with a foreordained conclusion laudatory of Western accomplishments, tribal peoples were given a marginal status as human beings.

For American Indians, the struggle of this century has been to emerge from the heavy burden of anthropological definitions that have made Indian communities at times mere laboratories for political and social experiments. Indian advocates are often very bitterly attacked by scholars when they question these experiments and articulate their own ideas which clash with accepted orthodox and comfortable interpretations about tribal people developed by academics. Indeed, some scholars become very competitive with Indians, believing that because they have studied an Indian tribe they therefore know more than any of the tribal members. The recent restrictions placed on anthropological research and the passage of the repatriation law have finally brought a reduction in the rate of exploitation of Indians by scholars but have by no means eliminated it.

If cultural evolution has been unkind to non-Western human societies, physical evolution has been devastating because it is the framework within which cultural anthropology is *supposed* to make

sense. That is to say, when we examine and compare physical evolution and cultural evolution, we discover a fascination with body
forms, jawbones, skulls, and other morphological data that quickly
turns into an obsession with flaked tools and intense debate about
when the European ancestors came down from the trees. But then
we move from a concern about the structure, shape, and size of
minuscule skeletal remains to a discussion—without having
proven anything—about possible beliefs of these creatures. The
articulation of physical evolution ceases and gives way to social
and cultural concerns. Presumably we enter a time period in which
it is impossible to determine skeletal changes because of proximity to the present, but perhaps also it is more comfortable to discuss the culture, since the physical evidence keeps being altered
in fundamental ways.

It would be exciting and highly informative to have a list recording every time a "scientist" has provided a definitive line of human ancestry. We would discover an incredible variety of
interpretations that exist and have existed, that new "men" are almost always relegated to the branches of our species tree and never
seem to fall into line as direct ancestors. We would also discover
immense problems in explaining how feet designed to climb trees
were suddenly capable of sprinting over savannas and across plains
with some degree of efficiency. Thus, at times we have had the sequence Java/Peking/Piltdown/Heidelberg/Neanderthal/Cro-
Magnon, and then modern man. With the discrediting of Piltdown,
apparently a scientific prank by a youthful Teilhard de Chardin or
perhaps an effort by him to foul the evolutionary nest on behalf of
Mother Church, we have other sequences which hold today as scientific explanations primarily because of peer politeness rather
than scientific insights. And we are always finding surprisingly
modern bones where they should not be.

Michael Cremo and Richard Thompson, in their impressive
study of the anomalies in physical anthropology, *Forbidden Archeology*, suggest that the presently acceptable sequence, which
has been carefully arranged to support the interpretations of

authorities in the field, moves from *Homo habilis* (2 million years ago) to *Homo erectus* (1.5 million years ago) to *Homo sapiens* (200,000–300,000 years ago) to *Homo sapiens sapiens* (not more than 100,000 years ago). After this sequence, which is highly suspect, we enter the lists of Java, Peking, and so on. Before we adopt this fictional genealogy, however, we should note that, as they explain,

> ... fossil skeleton remains indistinguishable from those of fully modern humans have been found in Pliocene, Miocene, and even Eocene and earlier geological contexts.

And they further ask us to consider that

> ... humans living today make implements not much different from those taken from Miocene beds in France and elsewhere ...[1]

suggesting that all the fuss and feathers being ruffled with respect to scrape tools, spear points, and other implements actually give little or no reliable information as to age and probable cultural context. *They are believed to provide accurate information because scholars have agreed to interpret them as if they did.* In fact, unknown to most of us who uncritically receive scientific studies as gospel:

> ... at most paleoanthropological sites, no hominid bones are found. The artifacts at these sites are attributed to *Homo habilis, Homo erectus,* the Neanderthals or *Homo sapiens* on the basis of their presumed age or their level of workmanship. ... [M]any Early or Middle Pleistocene sites currently identified with *Homo erectus,* for example, could just as well be identified with anatomically modern *Homo sapiens.*[2]

Human evolution, at least the evidence for human evolution, may exist more firmly in the minds of academics than in any location on Earth. We have been so well trained to accept uncritically anything that anyone alleging to be a "scientist" tells us that we do not really know what kind of evidence exists supporting evolutionary doctrines. William Fix, in his critical study of evolutionary

archaeologists, points out what was *actually* found at the limestone cave at Chouloutien, China, which became Peking man. The
skulls of some forty individuals, primarily women and children,
with their skulls bashed in represent Peking man. Fix writes,

> The Archaeological character of the deposits suggested the Pithecan-
> thropines were merely an item of somebody else's menu. The Anatomi-
> cal evidence, however, was such that in certain of their features these
> creatures appeared to be ideal candidates for a missing link between
> man and the apes—especially if one was predisposed to see this con-
> nection in the first place.[3]

Other evidence is equally as slim. In fact, if we took the fossils
and bones from all of the supposed human ancestors and put them
in a box, we would not need a very large box. We would be looking
at a few jawbones, some femurs, and certainly not enough evidence to either indict or convict. Fix can even be playful when he
describes the evidence for Ramapithecus:

> Fossils attributed to Ramapithecus have been found in India, China,
> Europe, and East Africa. Assuming these attributions are correct, it
> would seem Ramapithecus was widespread. Until recently, all that had
> been recovered of this creature were bone fragments of its face and jaw.
> We have no idea if it walked on two or four legs, or whether it was hair-
> less, sported a sleek black pelt, or was covered with a light purple fuzz.[4]

In all honesty, therefore, "science" should drop the pretense of
having absolute authority with regard to human origins and begin looking for some other kind of explanation that would include
the traditions and memories of non-Western peoples.

The question of human origins is critically important to the
Natives of North America, not simply as establishing a linkage of the
red race to the other branches of the human species but as providing
a clear and sensible picture of how our species did originate. *The
problem basically is that no Neanderthal skeletal remains have been
found in the Western Hemisphere. And no traces of the other ancient creatures have been found.* Since American anthropologists

and archaeologists are committed to supporting an outdated interpretation of human origins that sees Neanderthal as a predecessor to Cro-Magnon, that can only mean that American Indians are later comers to this hemisphere, having had to wait (at least in the minds of scholars) until Neanderthal evolved into Cro-Magnon and then for a convenient ice age when the North American continent could be linked with Asia.

Equally bizarre is a theory which seeks to justify the late classification of American Indians in the evolutionary scheme, cited by William S. Laughlin in his essay on the Bering Strait:

> Assuming that skin pigmentation is adaptive and responds to sunlight, then the aborigines in the equatorial regions of America should be as heavily pigmented as those in Africa and other regions of heavy pigmentation. That this is not the case has been used as an indication that the Indians have been in the New World for a period not much longer than 10,000 years [Haldane, citation omitted].[5]

What exactly does that kind of nonsense mean? Theoretically, all people should have reasonably similar skin pigmentation if human evolution had any basis at all, since all societies must originate from whatever pre-*Homo sapiens* we can identify as ancestral. Are scientists suggesting that skin pigmentation fades dramatically if people live in a more temperate climate or wear clothes? Some Indian tribes are darker than others. Can we arrange the relative time of arrival in the New World based on suntans? I suppose this theory does suggest that white-skinned people originated in the polar regions—an idea that would be rejected immediately by the scholars who prefer the suntan theory of classifying human societies.

How do we set this most powerful group of scholars to look beyond doctrine and begin to consider alternative interpretations? Many discoveries of early man, in fully modern form, have been made in the Western Hemisphere, but they are dated very late because of doctrinal considerations, not because of the evidence gathered by scientific excavations. Cremo and Thompson discuss the resistance of most scholars to admitting any early date for human occupation in the New World and note:

> The Sandia Cave discoveries, along with the finds made at Hueyatlaco, Calico, and Toca da Esperanca, strongly suggest a human presence over 200,000 years ago in the Americas. This challenges not only the orthodox time estimate for the entry of *Homo sapiens* into North America (12,000 years ago) but also the whole picture of human evolution, which has *Homo sapiens* arising from *Homo erectus* in Africa about 100,000 years ago.[6]

Many European scholars would not be frightened by such a change, and in fact many welcome it and are puzzled at the apparent lack of interest among American scholars.

Werner Muller, one of the late giants in European anthropology, in his startling book *America: The New World or the Old?*, describes the progress which has been made in recent times in defining the human ancestral tree.

> Human paleontology has drifted away from the idea of deriving sapiens genetically from earlier Homo forms; neither Neanderthal nor Heidelberg man are considered part of the family tree of present-day racial types on the European continent. Instead it is becoming more and more accepted to place Stone Age human types chronologically parallel to each other rather than in succession and to ascribe a clearly developed form to *Homo sapiens* as early as the beginning of the Quaternary.[7]

We are most probably really talking about the emergence of several types of modern men on different continents. In early January 1997 an international team of anthropologists announced a redating of twelve Java man skulls that placed their age between 27,000 and 50,000 years instead of the former dating of hundreds of thousands of years. This new dating meant that *Homo erectus* and *Homo sapiens* were probably contemporaries instead of separate steps in the human evolutionary line of descent.[8] We are still at a loss to explain human origins. But the absence of Neander-thal and the presence of Cro-Magnon in the Western Hemisphere appear to simply be *a fact* of human existence on the planet and *not* an indication of a late entry via a land bridge at all.

Suffice it to say that for more than a century American anthro-pologists have held firmly to the latest possible dates they could justify when describing American Indian tenure here. John Alsozatai-Petheo, a European anthropologist, described the mind-set of American anthropology well:

> For ... decades, American anthropologists would labor under the view of man's relative recency in the New World, while the mere mention of the possibility of greater antiquity was tantamount to professional sui-cide. Given this orientation, it is not surprising that when the evidence of the antiquity of man in America was finally reported from Folsom, Clovis, and other High Plains sites, it was rejected out of hand by estab-lished authorities despite the clear nature of the evidence at multiple locations, uncovered by different researchers, and seen and attested to by a large variety of professional visitor/observers.[9]

It is not that verification of Folsom and Clovis cultures by the discovery of their points and tools was rare and therefore speci-mens had to be carefully studied. A friend of mine, Bill Barker, a long-standing Denver radio and newspaper personality of the postwar years, found several points one evening while walking around at The Fort restaurant after dinner. So the evidence was overwhelming and abundant—if people wished to acknowledge it. Today, similar evidence of long-standing Indian occupation, and perhaps even origin, in the Western Hemisphere is emerging—only to face the same stubborn and outdated reception given Folsom and Clovis. Claude Levi-Strauss, perhaps the foremost anthropolo-gist in the world, complained in an article in *The New York Review of Books* in 1995 that "... here and there in the Southern Hemi-sphere, and more specifically in Brazil, settlements far more an-cient than that [10,000 B.C.] on the order of thirty or forty thousand years, have been ascertained by carbon-14 dating."[10] Cremo and

Thompson sum up what has *really* been happening in American and Western Hemisphere archaeology that is hardly known outside the profession:

> In the 1960s, highly sophisticated stone tools rivaling the best work of Cro-Magnon man in Europe were unearthed by Juan Armenta Camacho and Cynthia Irwin-Williams at Hueyatlaco, near Valsequillo, 75 miles southeast of Mexico City. Stone tools of a somewhat cruder nature were found at the nearby site of El Horno. At both the Hueyatlaco and El Horno sites, the stratigraphic location of the implements does not seem in doubt. However, these artifacts do have a very controversial feature: a team of geologists working for the U.S. Geographical Survey gave them dates of about 250,000 B.P.[11]

This report is one of the few times that I will try to believe a government agency's results.

Cremo and Thompson record a couple of instances in which the discoverer of an early site has been discredited—indeed, when even his friends and potential supporters have lost jobs and prestige for finding early sites, indicating that science and the academy are everything but impartial examiners of facts and purveyors of truth. The most blatant incident concerned Dr. Thomas Lee of Canada. Excavations were made at a site in Sheguiandah, Canada, between 1951 and 1955 by Lee, an anthropologist working at the National Museum of Canada. Preliminary evidence indicated the site might be between 30,000 and 100,000 years old.

The evidence not only conflicted with accepted doctrine, it would have made it necessary to revise estimates of the stages of North American glaciation. The scientific establishment went after Lee. He lost his position at the museum and some of his papers on the discovery were "lost." "I was hounded from my Canadian government position by certain American citizens on both sides of the border and driven into eight long years of blacklisting and enforced unemployment," Lee wrote.[12] A prominent anthropologist who visited the site, after expressing horror at the discoveries, advised Lee to fill in the excavations and build his reputation on work that was much safer doctrinally. Thus does science march forward.

Two scholars stand out, however, as people who have made a good-faith effort to raise the question of early American Indian occupancy here. Let us briefly look at their efforts and see what transpired. In the February 1963 issue of *Current Anthropology*, E. F. Greenman of the University of Michigan published an article entitled "The Upper Paleolithic and the New World," which argued on the basis of a number of categories of evidence that human beings might have crossed into the Western Hemisphere, particularly eastern North America, sometime during an interglacial period.

Fellow scholars were given advance copies of the article and asked to respond to Greenman, and he was given copies of their remarks and allowed to frame rejoinders to some of the points. By American standards, where heresy can produce dismissal from a job, boycotts of one's publisher, and accusations of mental illness for daring to think unusual thoughts, the exchange is moderately civilized. Since Greenman had published at the end of his academic career, there does not seem to have been retaliation against him.

In the technical discussion, some critics did score points when they argued that Greenman's examples were taken from such widely varying sources that a good comparison, at least one good enough to carry the argument, was not made. It is interesting, however, to note some of the arguments Greenman initially made and how his critics responded. His best arguments, in my mind, were three: that the northeastern United States has a significant number of artifacts of early man while the Bering Strait has virtually nothing; that canoes of the Beothuk peoples, now extinct, were designed for use in open water more than on eastern rivers; and that there is some similarity in cave drawings.

Although use of extensive quotes is often tedious, I am reproducing Greenman's arguments on two specific points with

excerpts from the critics' responses because they are so revealing of the determination of scholars to defend orthodox theories
to the detriment of the facts.

> [Greenman argues:] ... there is no significant accumulation of Pale-
> olithic tools or other traits on either side of Bering Strait, and the evi-
> dence is no better today than 30 years ago. On the other hand, looking
> to the eastern side of the continent, Upper Paleolithic traits are found
> clustered on opposite sides of the North Atlantic, in the Biscayan area
> and in Newfoundland.

> [T. Van der Hammen responds:] ... we should be very careful in draw-
> ing definite conclusions from the fact that *little or nothing is* known
> from the Bering Strait area and the adjoining area, Northeastern Asia,
> in connection with the eventual migration of late-Pleistocene man by
> that route. The region has not been extensively investigated and if we
> compare it with the development of our knowledge of Paleo-Indian
> cultures in the United States, much may change within the next de-
> cades. [Emphasis added.]

> [A. D. Krieger responds:] The alleged failure of the eastern Siberian
> and Alaskan areas to produce evidences of the earliest migrations into
> America cannot be taken too seriously. While this kind of evidence is
> rare in the far northern regions, there are many geographical reasons
> for the difficulty in finding sites of great antiquity there, and some very
> important discoveries are being made.

> [Thor Heyerdahl responds:] Let us therefore not despair even if the
> future should deprive us of material evidence of Asiatic migrations into
> Northwest America across the tracts of land surrounding the Bering
> Sea; a supplementary passage in the same northern area is provided by
> the warm current permanently sweeping eastwardly off the shores of
> the Aleutian chain.

This nonsense, readers, passes for scholarly discourse. Van der
Hammen cautions that even though we have no evidence, the situation may change radically in the next decades; Krieger says that
"some very important discoveries are being made"—apparently
without the knowledge of Van der Hammen and others. Finally,

Heyerdahl, while agreeing that a North Atlantic passage was feasible, argues that it doesn't matter—he can prove that Indians came over in boats on the Japanese current, assuming it was functioning in those days.

Let us consider another point. The North Atlantic coast does not offer any greater opportunity for study, and yet we do have a multitude of artifacts found there. We do have considerable knowledge about the Eskimos, and that should give us some indication of life in this particular area so that we can project by analogy what kind of conditions must have existed to make a migration reasonable. But the Eskimo evidence does not support the Bering Strait theory.

> [Greenman notes:] It is the same kind of thinking that has derived Eskimo from Siberia even though 98% of it is in North America. There has never been any more indication that the traffic across Bering Strait was from west to east than there has been of any significant traffic at all. I believe there was nothing in this area before about 6,000 years ago, and that after then the movement was, for a few centuries at least, from east to west.
>
> [Juan Schobinger argues:] It is almost certain that Eskimo culture, although developed in America, has Asiatic roots—the fact that today only 2% of the Eskimos live on the Siberian shore seems a rather ingenious objection ..."[13]

What can be said of this kind of argument? Perhaps we should note that since around 5 percent of the Swedes, French, Germans, Lithuanians, Japanese, Swiss, Irish, you name it, now live in the United States or even in Chicago, we can take it as scientific gospel that they must have originated there and migrated across the Bering Strait or across a mythical North Atlantic land bridge to what are now those countries. The birthplace of European culture may indeed be Skokie, Illinois.

A full reading of the critics would reveal typical scholarly rejoinders: lack of evidence does not mean we can't believe a thing if we wish; the future will show us to be right; already some very important but unknown things are being discovered; we have already linked this theory to several others of dubious foundation and if we disconnect it the whole field will collapse; most reputable scholars would not endorse such an idea; the methodology is faulty because the comparisons are not ones I like; and so forth into the night.

I do not think Greenman makes his point because it is too broad. He does prove that there is *some* connection because there appear to be similar traits and traces on both sides of the North Atlantic. But who went which way? Here I am not tied to Java/Peking/Neanderthal/Cro-Magnon and so I think it is an open and interesting question that needs more thought. Perhaps we should resurrect Piltdown and ask him.

As mentioned previously, Werner Muller provides considerably more thought about this problem in his book *America: The New World or the Old?* Briefly, Muller simply casts aside previous theories and begins with what we know about Pleistocene North America and the people who inhabited it. I was impressed on first reading the book but decided to check it out with anthropologists at the Smithsonian. Upon asking about Muller I was told that while he was a revered and generally competent scholar most of his life, toward the end he got a little crazy and published this book which varied considerably from what "responsible scholars" believe. Since in matters that deal with anthropology, and especially pre-Columbian America, the Smithsonian is more often wrong than Immanuel Kant was punctual, I take this description as a firm endorsement that Muller is right in most of his arguments.

Muller develops an excruciating argument for the very early origins of some Indian tribes based upon the style of architecture and the horizon astronomy and calendar recording. The scenario, as I now understand it, is that at least four different kinds of people lived in the far northern reaches of North America at a very early time, involvement with Neanderthal and Cro-Magnon being sidestepped for the moment, although it can be argued that these four groups are all Cro-Magnon peoples.

These peoples were the Salish, the Sioux, and the Algonkians, who presently live in North America, and a fourth group of mean-spirited, white-skinned, bearded people. The members of this last group were not very friendly with the three Indian groups and in fact may have caused them so much grief that the Indians decided to move south to get out of the way. But Muller hints strongly that some major kind of climatic catastrophe occurred (I suspect the ice ages but I doubt he would go this far) and the Indian groups spread down south in separate migrations, the Salish to the Pacific Coast, the Sioux to the Plains, and the Algonkians to the Great Lakes and eastern woodlands. The white-skinned people moved *eastward* across the North Atlantic into what is now Scandinavia and western Europe.

This scenario fits exceedingly well with what we know about the populating of western Europe. We are always reading that with the decline of the glaciers, Cro-Magnons entered northern and western Europe and routed the Neanderthals. We do know that the Cro-Magnon probably entered Europe from the west—at least the best sites for the Cro-Magnon are on the western shores of that continent, and if they came from the east or south they apparently suppressed their desire to paint in caves until they reached southern France and Spain.

Not much effort has been made to correlate all of the traditions of these Indian groups with Muller's theory, but he does cite the Cheyenne and Sioux to the effect that they once lived in the Far North and were pushed south by a major climatic catastrophe. Most startling for me, and very supportive of Muller's thesis, is an old Salish story I located to the effect that they were once enemies with the Sioux at a remote time. Between the Salish and Sioux we find the Crow, Blackfeet, and other tribes who should have been, and in recent historic times were, enemies of the Sioux. The Salish story does not seem to record the presence of these other tribes, and certainly if war parties from either Sioux or Salish were looking for a free-for-all, they could have been adequately accommodated by either Crow or Blackfeet very easily.

In spite of this radical reorientation of origins, Muller remains conservative on his North American dating, finding the major part of the displacement and European migration at around 42,000 B.C. with the onset of the interstadial of the Wurm glaciation. He can therefore accept the dating of the La Jolla fossils at 48,000 B.C. without hesitation. He does cite a very controversial bit of evidence for early Cro-Magnons/Paleo-Indians that many scientists may wish to avoid. Declaring that American paleoanthropology has always been a victim of misfortune, and implying that it has never been a worldwide mainstream participant, Muller revives discussion of the Calaveras skull, pointing out that the first discoveries were made prior to Darwin's theory, basically by gold miners who called in the geological experts of the day to verify their discoveries. And he says:

> The most important find from this time is the Calaveras skull discovered in February 1866 during the digging of a shaft in a gold mine near Altaville in Calaveras County, California. The cranium was found at a depth of 43 m., imbedded in a gravel layer under volcanic tuff which came from the Tertiary eruptions in the Sierra Nevada. No scientist of the time questioned the credibility of the frequently re-examined circumstances surrounding the find; it was not until the emergence of evolutionary dogmas that the find sank into oblivion.[14]

Muller then discussed all manner of very early remains found in America that were discredited because they did not fit into the evolutionary scheme. But nothing is as spectacular as the human skeleton found beneath old lava deposits, since this fact/artifact calls into question the geological time scale itself. I shall deal specifically with that issue in chapter 8.

Tribal traditions themselves vary considerably regarding the origin of man. Much of this knowledge is esoteric, revealed in a ceremonial setting and dependent upon some rather surprising and heretical views of time, space, matter, and cosmic purpose. My goal here is not particularly to support any specific tribal tradition but simply to examine the situations in which American

Indians find themselves, when confronted by the scientific mind that dismisses us out of hand. This dismissal is primarily based on doctrinal premises, not on a body of evidence that provides proof of scientific claims.

If we surrender the old evolutionary framework for the human species, admitting that we have only small bits of bone and some flaking and chipping tools which could have been made at any time, including today, we basically are dealing with the two types of modern man—Neanderthal and Cro-Magnon. Neanderthal simply does not appear in the Western Hemisphere and seems to have been driven from western Europe by invading Cro-Magnons during a glacial interstadial. *That scenario is really all we reliably know at the present time.* We should begin our investigation of human origins at that point and honestly look at the actual materials we do possess so as to create a more realistic interpretation of the history and experiences of our species. Until then, the political implications of classifying American Indians as some group of late-developing Cro-Magnon creatures stand in the way of equal treatment of Indians today.

4

Low Bridge—
Everybody Cross

IT MAY APPEAR THAT I HAVE SUFFICIENTLY DISCUSSED the origins of man and thereby eliminated the Bering Strait theory as a possible explanation of the source of the occupancy of the Western Hemisphere by American Indians and that devoting more time to this idea is superfluous. Nothing could be farther from the truth. Most Americans do not see the connection between the different scientific theories, nor do they understand that a shift or collapse of a major scientific doctrine requires a significant adjustment of all subsidiary doctrines that relied on it for their validity. Thus, people accepting the idea that outmoded explanations of human evolution have been modified substantially will continue to hold with the Bering Strait theory even though to do so is a great inconsistency. But another point must be made which requires a chapter of discussion—and that is whether or not the Bering Strait is simply shorthand scientific language for "I don't know, but it sounds good and no one will check."

There are immense contemporary political implications to this theory which make it difficult for many people to surrender. Considerable residual guilt remains over the manner in which the Western Hemisphere was invaded and settled by Europeans. Five centuries of brutality lie uneasily on the conscience, and consequently two beliefs have arisen which are used to explain away this dreadful history. People want to believe that the Western Hemisphere, and more particularly North America, was a vacant, unexploited, fertile land waiting to be put under cultivation according to God's holy dictates. As Woody Guthrie put it: "This Land is your land, this Land is my land." The hemisphere thus belonged to whoever was able to rescue it from its wilderness state.

Coupled with this belief is the idea that American Indians were not original inhabitants of the Western Hemisphere but latecomers who had barely unpacked before Columbus came knocking on the door. If Indians had arrived only a few centuries earlier, they had no *real* claim to land that could not be swept away by European discovery. Aleš Hrdlička of the Smithsonian devoted his life to the discrediting of any early occupancy of North America and a whole generation of scholars, fearfully following the master, rejected the claims of their peers rather than offend this powerful scholar. Finally, the embarrassing discovery that Clovis and Folsom points abounded in the western states forced the admission that the Indians might have beaten Columbus by quite a few centuries.

These ideas have great impact on how non-Indians view the claims for justice made by Indians. A personal experience may illuminate the impact of the Bering Strait on Indian rights. After Wounded Knee II in 1973, there were a number of trials of the people who had occupied the little village on the Pine Ridge Reservation in South Dakota. Each defendant had as his or her first affirmative defense to the criminal charges filed against them an avowed belief that the 1868 Fort Laramie treaty was still valid and that the protest was justified as a means of forcing the United States to live up to the terms of the treaty. This defense was then taken from every case and consolidated as one hearing in Lincoln, Nebraska, which dealt solely with this argument. Had the Indians prevailed in this contention, all the trials would have been rendered moot.

Much evidence was given at Lincoln concerning the relative state of civilized life at the time the treaty was made. The cultural achievements of the Sioux Indians were recited in an effort to demonstrate that, for many purposes, but chiefly for the trial, the Sioux had a clearly defined culture, government, religion, and economics and should have been entitled to the respect and benefits which larger nations enjoyed. In legal terminology, the contention was that the Sioux, in making treaties with the United States, had entered into a protectorate relationship comparable in every way to that enjoyed by Monaco and Liechtenstein with larger nations in Europe. This kind of relationship would then void the widely held belief that Indian tribes were mere "wards" of the government, as a confused portion of the John Marshall *Cherokee* cases had said.

Several traditional people did not want evidence on the Bering Strait offered because they preferred to rely on their own view of how the Sioux people had come to be. Some wanted to talk about an origin from an underground world near Wind Cave, South Dakota; others thought that stories about living in or near the Gulf of Mexico would be sufficient; and still others wanted to discuss the stories about living in the Far North, traditions that Werner Muller had used in his new theory of the human occupancy of North America. None of these accounts would have been understood in a Nebraska courtroom no matter how sympathetic the judge because they varied considerably with scientific beliefs about the Bering Strait. So some discussion was presented on the Bering Strait.

I was standing in the hallway of the courthouse smoking a Pall Mall (in those wonderful days when you and not your peers chose your vices) and a lady approached me all agiggle about what had taken place that morning. She gushed over what had been said about the Bering Strait as if she were the chairperson of an anthropology department and left me with the comment: "Well, dearie, we are all immigrants from somewhere." After reflecting on her comment for a moment, I wanted to run down the hallway after her and say, "Yes, indeed, but it makes one helluva difference whether we came 100,000 years ago or just out of boat steerage a generation back."

Her remark was symptomatic of the non-Indian response to the pleas of Indians. By making us immigrants to North America they are able to deny the fact that we were the full, complete, and total owners of this continent. They are able to see us simply as earlier interlopers and therefore throw back at us the accusation that we had simply *found* North America a little earlier than they had. On that basis, I would suppose, no nation actually *owns* the land its citizens live on, with the exception, if we accept early archaeological findings, of the people of Africa, where human evolution is believed to have begun.

In the 1960s, a group of California Indians protested at an Indian Claims Commission field hearing against a ruling that the California claims would be consolidated into one complaint, instead of allowing the individual tribal groups to file specific claims for their lands. The exchange between the Indian protestors and Chief Claims Commissioner Arthur V. Watkins got very heated at times. Watkins was a former U.S. senator and his anti-Indian sentiments were well known. He had introduced the termination policy in Congress during the 1950s (to dismantle reservations and relocate Indians to cities) and was rewarded, after he had lost his Senate seat, with an appointment to the Indian Claims Commission where he could do further damage to Indians. At one point, Watkins screamed at the Indians: "Go back where you came from," implying that they had recently traversed the Bering land bridge, perhaps during the Great Depression, and should go back to Asia.

Most scholars today simply begin with the *assumption* that the Bering Strait migration doctrine was proved a long time ago and there is no need to plow familiar ground. Jesse D. Jennings and Edward Norbeck's *Prehistoric Man in the New World* provides a compendium of papers discussing the state of research and field investigations dealing with the earliest sites of human occupation in the Americas. The introductory article has a single sentence on the Bering Strait and the essays proceed without the slightest doubt that they are being built on a strong foundation. Since these scholars were so confident of the validity of the land bridge doctrine I assumed that there was, somewhere in scholarly publications, a detailed article which cited evidence and arguments that proved, beyond a reasonable doubt, that Paleo-Indians had at one time crossed from Asia into the Western Hemisphere. I was unable to find anything of this nature.

I did locate a splendid book entitled *The Bering Land Bridge*, edited by David M. Hopkins, that appeared to be the answer to my inquiry. Alas, most of the articles dealt with technical geological and meteorological theories having nothing to do with human migrations. Only two articles even hinted at a discussion of migrations over the strait. H. Muller-Beck wrote that it had been

> ... established conclusively that glaciers flowing from the Canadian shield coalesced with those originating in the Rocky Mountains during some part of the Wisconsin glaciation: this coalescence may have lasted from as early as 23,000 until as late as 13,000 years ago. During most of this interval, when Alaska was connected with Siberia by a wide Bering land bridge, an ice barrier would have separated Alaska from central North America and contact between Alaska and central North America would have been extremely difficult for land animals and man.[1]

Muller-Beck also stated that what scientists were interested in was

> ... the diffusion of technological traits rather than population migrations in themselves: population movements are difficult to trace and have little relevance to the present problem.[2]

This clarification was another puzzle, since his article was entitled "Migrations of Hunters on the Land Bridge in the Upper Pleistocene" and it seemed likely that population migrations would be important to the topic. Did "traits" migrate without people?

The second article was "Human Migration and Permanent Occupation in the Bering Sea Area" by William Laughlin. Laughlin had graciously come to the Wounded Knee trials in Lincoln to discuss migration across the Bering Strait, so I looked forward to reading his article. But this article was devoted largely to a discussion of the Aleutian Islands, whose inhabitants he views as "quite distinct" from American Indians. No evidence was cited to show that scholars had proven that Paleo-Indians, or any other kind of Indians had traversed the Bering Strait at any time. Describing the land bridge, Laughlin painted a dismal picture:

> The interior landscape was evidently a low rolling plain, for the most part devoid of relief, studded with bogs and swamps, frozen much of the time, and lacking in trees or even many bushes. Grass-eating herbivores may have been present in fair numbers. The human adaptation to this region must surely have been that of big-game hunters, living by means of scavenging dead mammoths and such bovids as caribou, bison, and musk-ox, and by intentionally hunting live animals.[3]

But even this boggy, swampy land was not conducive to human migration. Laughlin pointed out that:

> Conditions in the interior [of Alaska] were severe, and likely only a few of its inhabitants found their way into North America; these wanderers probably became the ancestors of American Indians.[4]

Notice that Laughlin does not say for certain that any of these inhabitants crossed the Bering Strait—he only says it was "likely" that a few people did. We get no evidence at all that any Paleo-Indians were within a thousand miles of Alaska during this time. No sites, trails, or signs of habitation are cited. And that is it— Laughlin is the acknowledged dean of American Bering Strait scholars, and he offered no concrete evidence whatsoever to cite in support of this theory. I must conclude that generations of scholars, following the so-called scientific method of inquiry, have simply accepted this idea at face value on faith alone. Here is more evidence that science is simply a secular but very powerful religion.

Scholars and popular science writers, in discussing the Bering Strait doctrine, usually do not address the many real difficulties which this idea presents. They reach a point where they must sound intelligent to their peers and readers and promptly spin out a tale of stalwart hunters trekking across frozen tundra or frolicking in suddenly warm Arctic meadows, and continue with their narrative. Looking at a map of the world, the proximity of Asia and Alaska seems too obvious to reject, but only rarely do scholars look at the map closely enough to see the absurdity of their claim.

We will look at two major geographical factors—the actual topography of eastern Asia and western North America—and the barrier presented by the Ice Age, since scholars insist that the Indian migration occurred during a warm period of one of the Pleistocene ice ages. We will then look outside the topical area of Indian studies to see if and how other scholars use the Bering Strait in their work.

Presumably, the Paleo-Indians are living somewhere in eastern Siberia, having migrated there millennia ago. We will begin their journey with hunting bands living along the Kolyma River, at least half of which lies above the Arctic Circle. Looking eastward they would find two formidable mountain ranges, the Khrebet Gydan and the Chukotskoye Nagor'ye, blocking their migration to the east. If and when they surmount these mountains and find their way to the shores of the eastern tip of Siberia, they must cross over the strait, and here most scholars insist that it was not a strait but a broad plain because the water that would have ordinarily covered it was locked up in the glacial sheet that covers North America in the eastern part of the continent. We will allow them to cross, whatever the conditions.

Reaching the area we know as present-day Alaska, the people encounter a forbidding set of mountains both above and below the Arctic Circle. The Baird, Schawat, Endicott, and Shublik chains face them on the north, the Kaiyuh and Kuskokwim Mountains are to the south, and on reaching the Canadian border they meet the Richardson Mountains and the continental divide of the northernmost chain of the Rocky Mountain group. To the south also are the Ogilvie Mountains and then the massive MacKenzie mountain chain with the smaller Franklin Mountains yet to the east. Finally, the hunters are out on a reasonably flat plain, although one that is not calculated to present a paradise for hunters, since it is, according to many scholars, covered with a thick glacial sheet.

It is theoretically possible for a group of humans, determined to relocate, to push through a seemingly unending set of mountain ranges to reach another location; the question is whether or not this migration really happened. A good practice in testing a theory is to find out what scholars say about a subject when they are discussing another topic and simply mention it as a peripheral part of their discussion of another area. If we were to ask Bering Strait advocates if there were people in Siberia during the glacial interstadials, a time when it was possible for people to move without freezing to death or falling into glaciers, we might be assured that the shores of Siberia were teeming with impatient hunters. Indeed, didn't 98 percent of the Eskimos move from Siberia to Alaska at some point? But suppose we just ask about life in Siberia at this time. Let us see what scholars say about Siberia when they are not addressing the Bering Strait theory.

Kazimierz Kowalski and N. K. Vereshchagin are two important European scholars specializing in eastern Europe and Russia, including Siberia, in the Pleistocene period. Discussing whether or not Paleo hunters destroyed the mammoth in Siberia, Kowalski wrote that the "... mammoth was probably never the principal game of human groups, and the traces of human colonization of Siberia at that time are very scarce."[5] Thus we are talking about a very small group of people even being in Siberia, let alone making the journey to Alaska. And these little groups were hardly a menace to the mammoth or any other megafauna. These few Paleo hunters were not just wandering around Siberia looking for an isthmus. Vereshchagin describes the kind of human occupation of Siberia that existed at the time when Paleo-Indians were supposed to be migrating:

> In the plains of eastern Europe and Siberia the life of primitive man was connected with river valleys. Large herds of mammoths, rhinoceros, roe, giant deer and reindeer, and boar roamed from south to north and back along the valleys and flood plains of the rivers. The inhabitants of steppe watersheds preferred meadows and forests of flood plains, especially in dry periods or when the ground was covered with ice crust, because then elk, bison, tur, horse, and even saiga and camel fed upon branches of bushes and trees.[6]

The handful of people who lived in Siberia at this time did not have migratory patterns west to east across mountain ranges and high plateaus. Rather, they spent their time moving from south to north and back again following game who seasonally grazed when the weather was decent. Since these people had more than enough game and so did not hunt any of the species to extinction, there was no good reason for them to pick up their things and begin moving into rugged mountain areas where hunting would be more difficult and grazing animals a minimal resource.

The Kolyma would be the last good river system into which these hunters would have moved. There may have been some temporary expeditions to see what the eastern lands looked like, but the chances that these people would leave good hunting grounds for poorer ones are slim. Assuming, for the moment, that groups of hunters were able to get to Alaska, how would they have fared? Here we will ask a geologist with no doctrine of migration to support. Stephen Taber, writing a geological paper in the *Geological Society of America Bulletin,* gratuitously commented on migrating Paleo-Indians:

> Early man would have had difficulty surviving in the nonglaciated areas of Alaska through the first period of deep freezing, and he could not have migrated southward across the ice barrier. During the epoch of deep thawing, conditions were more favorable for the existence of man in Alaska and for his migration southward than at any time since the first deep freezing of Pleistocene sediments; but this warm period was also a time of high sea level, when a land connection between Siberia and America is improbable; and the crossing of Bering Strait on ice is unlikely when the climate was warmer than it is now.[7]

We have only traced the most likely route and given scholars the benefit of the doubt by locating the Paleo-Indians on the Kolyma River in eastern Siberia. Jared Diamond, discussing the big-game hunter migration which he believes took place around 12,000 years ago, says that "the colonists [of Siberia] probably came from eastern Europe, where Stone Age hunters in what is now the

Ukraine built their houses out of neatly stacked bones of mammoths."[8]

If we locate the migrating Paleo-Indians in the Ukraine, then it is necessary to add about a dozen more mountain ranges and a goodly number of high desertlike plateaus, a considerable stretch of tundra, and no one knows how many other obstacles. The point that must be understood is that nobody really knows; they just seem to make it up as they go along. To suggest a Ukrainian origin for people who migrated across the Bering Strait in turn suggests that they had something definite in mind in wandering eastward, and that supposition cannot be substantiated at all. Almost every articulation of the Bering Strait theory is woefully deficient in providing a motive for the movement.

Let us now turn to the second great barrier to human migration over the Bering Strait—the Ice Age. In order to move Paleo-Indians across the Bering Strait we must have the water level of the ocean drop significantly so that the isthmus will be dry land across which they pass or, alternatively, wander. The Ice Age of North American glaciation has provided a wonderful explanation for most scholars who deal with this subject. They seem to manipulate the water level to whatever depth they need to support their narrative. I have heard of drops of 50 feet, of nearly 300 feet, and, at the maximum, of 500 feet. It all depends on how much land between Alaska and Siberia the scholar needs to prove the case. The water level must drop a minimum of 60 meters, or 200 feet, to have any kind of isthmus at all.

We do not know the causes of the ice ages. They can range from the sudden cooling of the sun, a shift in the poles, the solar system suddenly traveling through an area of intense cold in space, or even a cometary dump of water. Most scientists seem to believe that glaciation was a prolonged process of cooling that enabled a massive ice sheet to build in the Northern Hemisphere, and that as temperatures varied over a period of a million years, sometimes less, at least four stages of glaciation affected the Northern Hemisphere. Some scholars today are reducing the traditional four stages to only two, and a few adventurous souls are advocating only one real stage of glaciation with considerable variances in climate affecting the thickness and location of the major ice sheet.

The traditional mechanics of glaciation have snow remaining all summer and snow precipitating in increasing volume each winter until there is sufficient snow to become ice and produce glaciers. The actual mechanics of this process are suspect. We need both warming and cooling to an extent far in excess of what we can observe today in order to bring it together. The temperate zones and tropics must evaporate a substantial amount of water in the summer. Indeed, if we are going to drop the water level over 200 feet and create the Bering land bridge, we have to evaporate an incredible volume of ocean water, an estimated 20.82 million cubic kilometers, enough water to cover an area of 5 million square miles with a sheet of ice 1.2 miles thick. All this water must be evaporated in the temperate and tropic zones, either within a very short time, or a great deal more water, allowing for evaporation, over a considerable period of time, must be evaporated and put into moisture-laden clouds.

Assume that we do get warmer oceans and produce heavy humid clouds; we then have to move these clouds from the temperate and tropical zones in a northerly direction to get them to the latitudes where they can precipitate as snow. Moving moisture north is the most difficult part of the process because there seems to be a kind of natural "dew" line below which cloud humidity would fall as rain and above which it would more likely fall as snow. I would locate this line around the present border of the United States and Canada, although knowledgeable meteorologists might place it higher or lower. We cannot today conceive of a natural process that would evaporate this amount of water and transport it safely from the temperate zones to ensure that it precipitated as ice in Canada.

If a massive cometary "dump" did occur, bringing extraterrestrial water in massive amounts to the Earth, we might be able to trace the fall of the water, torrential rains in the temperate climates and horrendous cascades of ice in the northern and southern latitudes, making the Ice Age almost instantaneous. This scenario is described by Donald Patten in *The Biblical Flood and the Ice Epoch;* it makes a lot of sense and explains many different phenomena. We have always had the problem of locating the source of water in the freshwater Pleistocene lakes in Nevada, Lahontan, and Bonneville. Speculations suggest they were filled by glacial runoff from the Sierra Nevadas, but the magnitude of these lakes prohibits that explanation. So a quick dump of fresh water on the midcontinent might be a good way to fill those lakes.

An annoying technical problem is that on our planet the winds generally move from west to east and east to west depending on the latitude. Constant and reliable winds do not, as a rule, move from north to south or south to north. In the Great Plains we do have the occasional "Alberta Clipper," which brings freezing Arctic air down the east side of the Rocky Mountains and creates serious snow and blizzard conditions. We do have occasional winds from the south which bring unseasonably warm winds north to melt the snow. And there is the very strange "Chinook," or zephyr, wind that heats cold areas unseasonably for a few hours and vanishes. On the whole, however, our winds and weather do follow a west to east pattern in the geographical regions where we find the big glacial sheets, so we must deal specifically with that fact.

To get the Ice Age under way, then, we must violate almost all the present knowledge we have of how our winds, clouds, and humidity work and create a different scenario that has very warm oceans creating clouds which promptly move north, across the dew line, and dump incredible amounts of snow on the Great Plains, the Hudson Bay area, and western Europe. Strangely, as far as we can tell, the snow clouds do not affect Siberia and no glaciers are found there that remotely resemble what happens in North America and Europe. But there has to be incredible cold in Siberia at this time because we do find frozen mammoths and frozen tundra, and the "deep freeze," of which Stephen Taber wrote, really puts some very cold temperatures into the ground in these areas.

The Ice Age itself, as noted, has been broken up by many scholars into four or more stages, divided by periods called "interstadials," which means that some kind of warming process takes place in the midst of the glacial sheet. The onset of additional cold later then creates more glaciation. No one seems to have a good explanation why or how the weather warms and subsequently cools. People are satisfied simply to have these interstadials because they make it possible to explain why we find traces of human occupancy in some remote and obscure sites which should have been covered with hundreds of feet of ice. Scholars are also able to introduce the Cro-Magnon people into Europe during the Ice Age by manipulating the data of the interstadial. The scandal at Sheguiandah, Canada, comes about because much evidence seems to point to interstadial settlement of a location. Here, excavations, although performed under the most excruciating conditions, revealed significantly old human habitations and resulted in the dismissal of Dr. Thomas Lee.

Scholars also have invented strange concepts which they use to explain deposits which puzzle them. Thus, "advances" and "retreats" of glaciers are suggested to account for various kinds of gravels and clays whose presence would raise questions about the validity of a prolonged period of glaciation. A glacier apparently "advances" by moving southward and covering ground where it has not previously been located. By the same token, "retreats" would find the glaciers melting significantly at their southern edge to let a considerable tract of ground dry out and support plant and animal life. It is puzzling just exactly how glaciers do "advance." According to accepted ideas, the snow accumulates to as much as several miles high and, as pressures build, the glacier, or a significant part of it, then begins to stretch out over land which has not previously been affected by glaciation.

This movement of glaciers has always given me problems. It is always presented as if the planet were a Sherwin-Williams paint logo and the glaciers just naturally began to move south, so people have rarely asked how an inert sheet of ice can start to travel.

Present theory would have glacial arms moving away from the Chicago region and traveling southward to the St. Louis, Missouri, area, not an inconsiderable distance. Geologists have primarily studied Alpine, Alaskan, and Greenland glaciers and on that basis have developed the idea of moving glacial ice sheets. But all of these glaciers originate in the mountains. They thus have the assistance of gravity as they move downhill into the flats. It would not seem difficult to move ice downhill in a mountain area, because you would have a solid rock surface underneath the ice, melting waters to lubricate it, and gravity to occasionally coax it along. Melting water would also run downhill, making some movements of the glacier a spectacular leap forward and simply not providing any encouragement during cold seasons.

If we transport this mechanism to the plains and woodlands of Canada, we have a different picture. Ice sheets, no matter how high, are resting on topsoil, so water simply seeps into the ground and finds its way from under the glacier to unglaciated terrain. We have no assistance from gravity at all. Indeed, moving the ice from Hudson Bay to St. Louis would mean moving it uphill for many hundreds of feet of elevation. We can simply surmise that ice can and does go uphill if scientists want it to do so.

I have devoted an inordinate amount of attention to the mechanics of glaciation because it is necessary for the reader to see the magnitude of the problem which glaciation presents for our Ice Age Bering Strait immigrants. While eastern Siberia, mysteriously, is not glaciated, the Alaskan mountain chains are victims of the glaciation. Existing Alaskan glaciers may be remnants of the original Pleistocene glacial sheet. If we are successful in getting any of the Paleo-Indians across the Bering Strait, they will simply have to remain in the central Alaskan marshlands until the glacial age has subsided and it is safe to travel. Or so we would think. But scientists, being an inventive sort, are not content to leave the settlers alone in Alaska.

The mechanism by which we move the Paleo-Indians from Alaska to the interior of the continent and then to the lower forty-eight is the "ice corridor." Jared Diamond brilliantly articulates this idea in an article in *Discover* magazine:

> At intervals during the ice age, a narrow ice-free, north-south corridor opened through this wall [the glacial ice sheet], just east of the Rockies. One such corridor closed about 18,000 years ago, when apparently there were not yet any people in Alaska. However, when the corridor next opened, 12,000 years ago, the hunters must have been ready, for their telltale stone tools appear soon thereafter, not only at the south end of the corridor near what's now Edmonton, Alberta, but also elsewhere south of the ice cap.[9]

We may certainly *need* an ice corridor if we are going to explain how the Paleo-Indians got through the North American glacial sheet, but does that make it a geological certainty or reality?

To be really useful, this corridor would have to extend not only along the east slope of the Canadian Rockies. It would have to extend clear into central Alaska so that people then living in those frozen marshlands would be inspired to see where it led. We cannot have Paleo-Indians forsaking the few flat areas where there is game and wandering across range after range of glaciated mountains searching for the corridor that leads them to the south. In addition, it would seem that snow clouds, as they do today in the Rockies, would hit the high mountains and then dump their snow first on the eastern slopes before transporting all that moisture clear across the Canadian plains to deposit it in the Canadian shield region. So we would have heavy snowfall precisely where scholars want a corridor.

American Indians, as a general rule, have aggressively opposed the Bering Strait migration doctrine because it does not reflect any of the memories or traditions passed down by the ancestors over many generations. Some tribes speak of transoceanic migrations in boats, the Hopis and Colvilles for example, and others speak of the experience of a creation, such as the Yakimas and other Pacific Northwest tribes. Some tribes even talk about migrations from other planets.

The Sioux, Salish, and Cheyenne remember their life in the Far North, which featured entirely different climatic conditions than we find today. The Sioux tradition, related by Thomas Tyon around the turn of the century, states:

> The seven council fires burned in a land where the trees were small and the leaves fell before the coming of each winter.
>
> The seven fires were lighted in a circle (the nations were camped together) and *Waziya* appeared in the council. He was a large man and clothed in heavy furs. He said, "Why do you stay here where the trees are small and the leaves fall? Come with me and I will show you where the trees grow tall and the leaves are green all winter."[10]

The Salish account has certain similarities. Ella Clark reports a tradition given to an interpreter in 1923 by four elderly Salish concerning Flathead Lake. To the question of origins, these old people said: "... the first Salish were driven down from the country of big ice mountains, where there were strange animals. Fierce people who were not Salish drove them south. So in our stories our people have said: 'The river of life, for us, heads in the north.' "[11] Since the memories of American Indians clash directly with scientific speculation, there is little room for compromise here.

Some tribal traditions do speak of ice and snow, which may be memories of North American glaciation, particularly since ice and snow are normal phenomena in the United States and remembering a really big snow would indicate that it was unusual. Most of these tales begin with the supposition that these groups were already present in North America prior to the onset of glaciation and quite possibly were observers of some of the climatic events of the Ice Age. The simplest ice tradition is that recorded by Julian Steward in a collection of Western Shoshone traditions but actually provided by a Northern Paiute person from Winnemucca, Nevada, concerning a large body of ice on the Snake River. Since it is short, it can be used to illustrate the casual nature of the account. It seems that Coyote took some of the Paiutes north to the Snake River:

Ice had formed ahead of them, and it reached all the way to the sky. The people could not cross it. It was too thick to break. A Raven flew up and struck the ice and cracked it [when he came down]. Coyote said, "these small people can't get across the ice." Another Raven flew up again and cracked the ice again. Coyote said, "try again, try again." Raven flew up again and broke the ice. The people ran across.[12]

Although there is some involvement with supernaturals, the basic story line is simply that the people went north, saw ice that went to the sky, and tried to cross it.

More complicated is the Chippewa creation story, which says that God tried four times to create the present world but the first three efforts were doomed to failure because there was too much ice. The fourth time the effort was successful. If this tradition is a memory of the four stages of North American glaciation, it implies that the glaciation occurred within a reasonably short period of time so that people remembered the process. Since the Chippewa flood story relates that the flooding was caused by rapidly melting ice, we might suggest that Chippewa traditions are something to be taken seriously.

The Hopi have a tradition that their clans had to make migrations around the Western Hemisphere at the beginning of this present world. Five clans—the Blue Flute, the Ghost or Fire, the Spider, the Snake, and the Sun—all migrated up the western side of the continent until they reached "… a land of perpetual snow and ice." Here they were tempted by Spider Woman to use their special powers to melt the mountains of ice and snow. Sotuknang, nephew of the Creator, then appeared and scolded them, pointing out that if they continued their activities they would melt the ice and snow and destroy the newly created world. They ceased their mischief but Spider Woman and her clan were punished by becoming the source of evil and discontent in the world.[13] It seems unlikely that the Hopi, living on the Colorado Plateau in northern Arizona, would be able to guess that the northern reaches of the continent were lands of perpetual ice and snow. This tradition must reflect a journey to the north.

The argument over the validity of the Bering Strait doctrine might continue on indefinitely, since most scholars are not inclined to take seriously the kinds of objections that skeptics raise. We need some additional evidence on one side or another to tip the scales and force a genuine reappraisal of the idea and the creation of a reasonable alternative. I did not realize how useful the doctrine was until I went outside the anthropological literature and began to look at how the scientists in other disciplines creatively used the Bering Strait for their own purposes. Pending the publication of all the "new" discoveries which Greenman's critics maintained were being made daily, let us now turn to the Bering Strait as it appears outside the Paleo-Indian context.

I was reading Donald Worster's popular historical study *Dust Bowl* one afternoon when I came across the following passage: "Horses and camels began their existence 45 million years ago in the North American grassland, migrating later across the Bering land-bridge to Asia. Bison followed the same route in reverse during the Ice Age and discovered a domain in which they could thrive."[14] I certainly wouldn't want to question or denigrate Worster's research or scholarly reputation, but it seems incredible to me that two grazing animals, terribly well adjusted to the grasslands of Kansas and Nebraska, could not thrive there and so would suddenly pick up their things and move north into an increasingly cold climate, where grass is at a minimum, looking for a better place to live.

I can't imagine thousands or perhaps millions of horses and camels struggling to get through the MacKenzie Mountains, or perhaps stampeding up Skagway Pass, crossing over the land bridge, and then being confronted with approximately a dozen rugged mountain ranges which they had to traverse before they found a home in the steppes of Asia. Did they suddenly change their diet, for the purposes of migration, from grass to tree bark and tundra or yearn for the Asian steppes in some mystical vision?

The bison migration, as some scientists tell it, has a lot more to offer in the way of credibility. I can imagine the scene. *Bison bison* and Mrs. Bison are peacefully grazing in central Asia without a care in the world when they look up and see horses and camels strolling by—the camels perhaps on their way to Egypt. Quick as a wink, *Bison bison* turns to Mrs. Bison and happily exclaims: "Honey, do you realize that there is an ecological niche for grazing animals now open in Kansas and Nebraska?" The whole herd is terribly excited at the prospect in spite of the fact that the monsters *Bison taylori* and *Bison latifrons* already graze most of the central American plains.

Word goes around central Asia and pretty soon the whole *Bison bison* species decides to cross the Bering Strait, knowing full well that a trail has already been made for them by the large herds of horses and camels that have previously made the crossing. The *Bison bison* are very pleased because Indians are also crossing the Bering Strait, and now everything is set for the great American West, 12,000 years hence, when the Plains Indians will hunt them and sell their hides and Buffalo Bill will achieve fame by nearly exterminating them.

Worster may have been a little enthusiastic about his date of 45 million years because Stephen Taber, examining Alaskan muck deposits containing animal remains, suggested that "near the beginning of the Pleistocene, elephants, bisons, goats, moose, wapiti, caribou, bears, wolves, foxes, and other mammals migrated from Asia to America, and horses and camels migrated from America to Asia."[15] It was not just the intuitive feeling that an ecological niche was open around North Platte, Nebraska, then, that encouraged the bison migration. It was the mammal fad of the day and any socially responsible species in Asia worth its salt was rushing toward the Bering Strait. We can surmise that horses and camels, watching the menagerie come loping across from Asia, decided to vacate North America while there was still time.

A short time later (in relative scientific terms), I found a book by L. Taylor Hansen entitled *The Ancient Atlantic* which combined orthodox scientific findings about this ocean with some strange anomalies to give a history of geological and human activities associated with the Atlantic. Hansen confirmed Worster's camel migration and added that "during the Oligocene the Aleutian bridge from Asia to the Americas was dry and functioning as a means for animals and plants to cross."[16] And, she said, "... while the Aleutian bridge was open in the Oligocene, some American species made their way to Asia. Among these was the baluchitherium, a giant rhinoceros measuring eighteen feet tall at the shoulder. He was therefore four feet higher than the Imperial mammoth and the largest land mammal. He crossed the Aleutian bridge into Asia, probably along with palm, oak and walnut forests of Canada."[17]

Now, I can see John Wayne, Rory Calhoun, and even Bob Hope and Bing Crosby struggling up Skagway Pass because they've done so in the movies. I can even, as a loyal admirer of science, try to visualize herds of horses and camels racing through frozen mountain passes in Alaska and Siberia. I cannot, however, imagine the largest land mammal who ever lived, four feet higher than the Imperial mammoth, moving by the thousands through the western Canadian mountains trying desperately to get across the Bering Strait to Asia before the sea level rose again.

Nor can I imagine forests of palm, oak, and walnut moving majestically west from Alaska to Siberia. I have great difficulty conceiving of their means of locomotion—other than the fantasy of scholars. Do you suppose that they "threw" their coconuts, acorns, and walnuts as far west as they could reach each fall and in stately procession marched right across the Bering Strait, putting their roots down in tundra and then continuing to lean westward each generation until the migration was complete? Minimally, this scenario would have required that the palms had previously left the Caribbean and Florida areas and moved into the Northwest Territories or come up the British Columbia shores in order to be in a position to take advantage of the land bridge when it finally appeared.

After I saw *Jurassic Park* and was thrilled to learn that most of the dinosaurs featured in the movie were in fact from the Cretaceous period, not the Jurassic, I was determined to learn more about them. Robert Bakker, who was the movie consultant, lives in Boulder, and so some friends called him to see if we could have lunch and learn more about the dinosaurs. He was always quite busy doing important dinosaur work and, considering the popularity of the movie, it was not difficult to see that he could not take the time to have lunch with fans, no matter how sincere they were.

I eventually did meet him and he is a splendid fellow. But while we were waiting week after week to meet him, I had a chance to purchase and read his excellent book *The Dinosaur Heresies.* Bakker apparently fancies himself as the enfant terrible of science and the great paleontological heretic, although his beliefs and theories are quite in line with Stephen Jay Gould and other popular apologists for orthodox science, so the title of the book is a little misleading. He is pretty much a party-line scholar with great personal energy and charisma.

Reading Bakker's book, however, I found chills running up and down my spine. Bakker knows dinosaurs and, while he identifies literally dozens of great-uncles, uncles, aunts, cousins, and shirttail relatives among the various species of dinosaur, he does not once identify any specific family trees which show evolutionary descent from one species to another. He does know dinosaur muscles, energy levels, diets, and environments intimately, so the book is well worth the reading. But what really attracted me about his discussion on dinosaurs was the fact that they also had crossed the Bering Strait land bridge.

My favorite passages are herewith reproduced. Giving us an elaborate description of *Protoceratops,* Bakker says that not one of these creatures "… has ever been reported from the rich beds of the American Judith and Laramie Deltas. Swampy meadows and broad humid floodplains were evidently not to Protoceratops's liking, though Canada and Montana did play host to relatives [see!]

in late Cretaceous times—the general Leptoceratops and Montanoceratops. Leptoceratops probably was an immigrant from Asia."[18] Bakker elaborates on this tantalizing hint by stating that "there were many advanced mammals and protoceratopsid dinosaurs in the Central Asian Highlands not found in Alberta, Montana, and Wyoming. But very late in the last epoch of the Cretaceous Period, the Asiatic mammals and dinosaurs began appearing in North America. These immigrants could only have passed over the Bering Land Bridge where the northeastern tip of Asia met America."[19]

Fortunately, Robert Bakker does not mince words. He suggests also that South America was once an isolated continent in which mammals and birds evolved into species found nowhere else. Then an isthmus was formed, probably due to the drop in water level of the seas during Pleistocene glaciation, which is now present-day Panama. With this connection made, North American fauna, in particular elephants, jaguars, deer, tapirs, and wolves, rushed into the southern continent. "These North American immigrants devastated the native fauna," Bakker maintains. "Most of the big South American species went extinct, victims of predation and competition from the Northerners, as well as of their diseases."[20]

Bakker is a respected scientist and so we should take his word on this matter, but it is difficult to believe that deer and tapir can eliminate giant ground sloths twenty feet tall, saber-toothed pouch mammals, and flightless killer birds larger than a lion merely by grazing areas occupied by these animals. Species, particularly grazing herbivores, generally can accommodate themselves to other grazers. Moreover, animals do not, as a rule, transmit diseases across species boundaries. The Panama land bridge is here invoked to explain events (the extinction of large mammals) which scientists cannot otherwise explain—and the explanation simply does not hold water.

Bakker is at least consistent with his arguments regarding land bridges. "The late Cretaceous world contained all the prerequisites for this kind of disaster," he writes. "The shallow oceans drained off and a series of extinctions ran through the saltwater world. A monumental immigration of Asian dinosaurs streamed into North America, while an equally grand migration of North American fauna moved into Asia."[21] From this description it seems likely that every time a narrow body of water was temporarily dry, hundreds if not thousands of species immediately dropped what they were doing and headed for the isthmus before it closed. Even Stephen Jay Gould is not above transporting animals over hypothetical necks of land if the occasion warrants. Witness his description of the Irish elk:

> The giant deer flourished in Ireland for only the briefest of times—during the so-called Allerod interstadial phase at the end of the last glaciation. This period, a minor warm phase between two colder epochs, lasted for about 1,000 years, from 12,000 to 11,000 years before the present. (The Irish Elk had migrated to Ireland during the previous glacial phase when lower sea levels established a connection between Ireland and continental Europe.)[22]

Since the interstadial was only 1,000 years long, or about the time between the fall of Rome and the discovery of America, the Irish elk must have been gathered on the shore waiting for the land bridge to open. It does not seem possible, considering the time that most scientists require for species to pass over a land bridge, for the large deer to make the transfer.

When reading these "scientific" explanations we must always remember that in order to have land bridges at all, or even an occasional isthmus, we are basically committed to moving a great deal of water around to create an ice age, or we are making the continents rise and fall a significant distance, or we are otherwise manipulating a monstrous amount of physical material just to make our theories and speculations seem reasonable. If scientists were required to solve these physical problems prior to their rather offhand remarks about migrations, there would be considerably fewer land bridges in scientific literature. Following orthodox methodology, we should not invoke activities of nature that we do not see operative today.

It occurred to me that I might be able to find an essay devoted solely to the question of the validity of land bridges, written when a scholar had no thesis of migrating species to defend and when the Bering Strait migration did not come to mind. And indeed such an essay exists. George Gaylord Simpson was about as close to a living deity in evolutionary biology as Mother Nature herself, and one day he sat down and penned a little piece entitled "Mammals and Land Bridges." We can assume that what was applicable to mammals might be profitably applied to dinosaurs and perhaps even to Paleo people.

Simpson uses a commonsense approach to the subject and suggests that only representatives of genera cross land bridges. A single genus does not by itself cross into new continents. More important, carnivores generally follow the herbivores they have been feasting on. "Where herbivores go, carnivores can and will accompany them, and carnivores cannot go where there are no herbivores. The postulation of land bridges on the basis of one or a few mammals is thus very uncertain. Unless there is a reasonable possibility that their companions have not been discovered, a theoretical bridge based on such evidence is probably unreal."[23] In other words, if we do want to move horses and camels to Asia and bison to America, we will probably want to ensure that carnivores accompanied them if we wish to make our case.

The objection raised earlier regarding human, and then mammal and dinosaur, expeditions across the Bering Strait—that the route had to traverse a set of rugged mountain ranges on both sides of the Bering Strait—is regarded by Simpson as a major barrier even if a land bridge does exist. "For many of these animals, such as the monkeys, the absence of necessary environmental conditions beyond the bridge is an evident reason for their stopping where they did. Others, like the bison, were evidently kept by analogous environmental barriers from reaching the bridge."[24] In other words, the bison simply would not have begun the tedious trip through the Siberian mountains, nor would horses and camels have tried to scale Skagway.

George Gaylord Simpson's conclusion, apparently unread or unheeded by several decades of scientific writers, is that "in the whole history of mammals there are exceedingly few cases (e.g., Lower Eocene between Europe and North America) where the evidence really warrants the inference of a wide-open corridor between two now distinct continental masses."[25] This conclusion supports Werner Muller's Canada to Scandinavia-England-France thesis and does not give much comfort to the myriad scholars who believe in the Bering Strait—for both animal and human migrations.

Not only does the more recent interpretation of human evolution militate against American Indians being latecomers to the Western Hemisphere, an examination of the Bering Strait doctrine suggests that such a journey would have been nearly impossible even if there had been hordes of Paleo-Indians trying to get across the hypothetical land bridge. It appears that not even animals or plants *really* crossed this mythical connection between Asia and North America. The Bering Strait exists and existed only in the minds of scientists.

Dr. Claude Levi-Strauss, in his article on Brasil, says that "... many archaeologists in the United States still subscribe to the dogma that this was the millennium [tenth] when human beings crossed the Bering Strait and set foot in America for the first time."[26] Levi-Strauss and many European scholars understand that there is no basis for the early dating of the so-called migration. Whether they have discarded the very idea of it is undetermined, but with the admission that various species of hominids coexisted, the way is being cleared for an honest examination of the question of origins on the basis of scientific investigation and not as a dogma that must be uncritically accepted.

5

Mythical Pleistocene Hit Men

SCIENTIFIC THEORIES ARE OFTEN BUILT on the most tenuous of evidential foundations and survive only because of the gentleman's agreement within scientific peer groups not to embarrass colleagues. One theory with dubious validity serves to provide the platform for articulating another theory which has even less to recommend itself and a third theory assumes that the first two are correct. By relying on outmoded general theories and doctrines, a scholar can skate out onto the pond of fictional enterprise, promulgate nonsense, and be taken seriously by his colleagues. Anthropology, archaeology, paleontology, and geology, in dealing with the Western Hemisphere, all rely on the idea that strata must be correlated according to certain preconceived time scales. Thus, any discoveries of human habitations prior to the Ice Age are usually rejected by archaeologists because they rely on anthropologists who in turn advocate the Bering Strait theory, in effect putting archaeology in a straitjacket.

Untangling the confusion is almost impossible because many subjects under discussion are so immense that most scholars must take the word of other scholars. There are no umpires to determine what is valid and what is not. But there are complaints about this situation. Alex Krieger, in an essay entitled "Early Man in the New World," described the situation as follows:

One of the greatest weaknesses of archeology is inadequacy of sampling. When new cultures are mentioned in print, there is rarely any attempt to discover how much field work was done to back up the claims, the amount of ground excavated, the number of artifacts found, or the portion of the site examined. Consequently, some extremely weak cases have been accepted and repeated in the literature, then repeated again by others until they become so firmly embedded in the minds of scholars that it is impossible to convince them that the original claim was based on no solid evidence.[1]

We have already seen that propensity both in the Bering land bridge and in the question of the human evolutionary sequence.

It seems nearly impossible, if we listen to Krieger, to correct the tendency of scholars to close their minds. Krieger's lament belies his claim that he understands the problem. In the same essay, five pages later, Krieger is discussing the archaeological discoveries at the Lewisville site northwest of Dallas, Texas. The excavators found a Clovis fluted point and conducted radiocarbon tests on associated vegetable materials found in Hearth No. 1. These tests produced dates greater than 37,000 years, were remeasured, and produced dates greater than 38,000 years—a real embarrassment to scientific doctrine because the makers of the Clovis points were not supposed to be in North America until around 15,000 years before the present. So the Lewisville site really marked a breakthrough in knowledge of the Pleistocene past of North America. But Krieger's own response to this discovery was the following:

> ... as I pointed out soon after discovery, there are alternative explanations, among them the distinct possibility that the Clovis point was planted in the hearth by someone not connected with the excavation or that, by some incredible accident, machinery used to excavate the huge borrow pit in which this and other hearths were exposed somehow caused the point to be dragged or lowered to this position.[2]

You will note that the Clovis point did not come under suspicion until the radiocarbon dating produced dates that were doctrinally unsatisfactory. So the early dating is rejected with the explanation that, unknown to the archaeologist, a prankster had dropped a Clovis point into the diggings when the scholars weren't looking—and in the alternative, that the scholars could not tell the difference between a recently planted point and something that had been buried there for some 38,000 years. But even another alternative exists: the excavating machinery was rarely cleaned while being moved from one site to another, so hanging on the backhoe were dozens of arrowheads, dinosaur bones, Clovis points, and perhaps some bottle caps—which no one noticed and which frequently fell into holes to produce wrong dates on vegetable materials.

With this background, aware of the hesitancy of orthodox scholars to accept new ideas and discoveries, we are ready to examine a theory—the "Pleistocene overkill" hypothesis—that has become popular in recent years among a small group of scientists and a surprisingly large segment of the public. Supporting this theory, and integral to its validity, are the outmoded sequence in which Neanderthal man precedes and evolves into Cro-Magnon man and the assumption that Paleo-Indians, the ancestors of today's American Indians, came trooping across the Bering Strait land bridge around 12,000 B.C., a belief I have suggested might be highly suspect and almost physically and culturally impossible.

The theory of Pleistocene overkill originated as an effort to explain the extinction or disappearance of a significant number of Pleistocene animals, most specifically the megafauna over fifty kilograms in weight, during or at the end of the Pleistocene era, a geological period which is in itself a complex and suspect classification. We thus move from the question of the validity of land bridges, which are sacrosanct for many scholars, to extinctions which are a great puzzle for most scholars, since geological strata from the first signs of life show sudden and mysterious terminations of planetary biota and fauna.

This "overkill" idea originated in a form approaching its contemporary expression when Carl Sauer, then one of the premier scholars on the environment, suggested more than half a century ago that Paleo-Indian hunters had killed all the big game animals which became extinct in the late Pleistocene era. Sauer thought that through the use of fire drives in which they both cleared large tracts of land for prairie grazing and eliminated the mammoth, mastodon, and a variety of other creatures, Indians had been responsible for the demise of the mammoth and mastodon. Sauer cited no large kill sites, mixed forest and plains animals indiscriminately, and could only point to the practices of some historic Indian tribes in burning grasses to encourage new growth the following year as evidence that Indians had even burned areas.

Sauer was easily routed in his efforts to gain acceptance for this thesis by Loren C. Eiseley, who wrote two articles raising questions that were never satisfactorily answered. Eiseley's criticisms were as follows:

1. Many kinds of smaller fauna also died during this period that could not have been killed by either fire drives or spears and atlatls—most specifically birds and freshwater creatures such as mollusks and frogs.
2. The larger species of bison, most probably *Bison taylori*, a giant version of our present *Bison bison*, became extinct along with many other large grazing animals but our bison, antelope, deer, elk, moose, and other familiar modern game animals did not.
3. Many animals were in fact forest dwellers and could not have been affected by grassland fires. It would have been impossible, given forest environments, to have exterminated whole species under any conditions.
4. There is no evidence, comparing modern examples of tribal hunting groups with *possible* ancient techniques, for any group to exterminate or even significantly affect an animal population unless the hunters and prey are restricted to a very small area.

5. Prey-predator ratios almost always adjust to conditions and
   do not get out of alignment unless there is a catastrophic
   decline of predators—the Grand Canyon Kaibab deer is
   always cited as an example of a lack of balance caused by loss
   by predators.

With fundamental criticisms such as these, the thesis was rejected by most scholars and was forgotten until it was revived in
another form by Paul Martin of the University of Arizona and his
colleagues in a 1967 book of essays entitled *Pleistocene Extinctions.*

Since these events, if they did indeed occur, happened some
12,000 to 15,000 years ago, why should it matter? It matters immensely because the image which science has given American
Indians is such that modern Indians are blamed for the extinction
of these creatures. Conservative newspaper columnists, right-wing
fanatics, sportsmen's groups, and scholars in general tend to see
the "overkill" hypothesis as symptomatic of a lack of moral fiber
and ethical concern for the Earth among Indians. Some people
are offended by the thought that many people believe that Indians were more concerned and thoughtful ecologists than modern
industrial users. Advocating the extinction theory is a good way to
support continued despoilation of the environment by suggesting that at *no* time were human beings careful of the lands upon
which they lived.

I can speak here from firsthand personal knowledge. In 1990, I
was invited to speak at Stanford University, trumpeted as the
"Harvard of the West," to celebrate its one hundredth anniversary.
I was asked to speak on the Indian relationship with the land, and
I tried as best I could to outline the philosophical principles I
thought would be meaningful to the audience and the values I
thought were involved in the Indian perspective on the natural
world. The first question from the audience when I finished was a
person asking whether I didn't think running hundreds of buffalo
over a cliff was wasteful. The tone of the question implied that the

previous weekend other invited Indian speakers and myself had destroyed hundreds of bison somewhere in Wyoming. Since the only recent slaughter of buffalo that I could remember was the Super Bowl, I took offense and refused to answer any more questions. I did not think that political correctness, applied retroactively to 15,000 B.C., was appropriate.

Lest the reader think I am overly sensitive, let us see what some of the respected science writers say about this thesis. They have, as a rule, accepted it wholeheartedly. Robert Bakker, whom we met in the previous chapter, casually noted that "... once they got going, our primordial forefathers cut a wide swath through both the Old World and the New, exterminating dozens of big species of mammal—mammoths, mastodons, sabre-toothed cats, giant ground sloths, to name but a few. And they killed tortoises."[3] Robert Ardrey, in *The Hunting Hypothesis*, waxes eloquently about the subject. "And so it is," he writes, "that every fossil record points to the probability that within a thousand years after our arrival in America across the land bridge we similarly exterminated the mammoths."[4] Writing in such absolute terms encourages the layperson to believe that these questions have been proven beyond doubt, when they never have been, and it is wholly irresponsible.

Ardrey is not talking about "we" in the generic human sense. "The new hunters, who would father the American Indian, left an unmistakable record: within a few thousand years they and their descendants, armed with only throwing spears and Asian sophistication, exterminated all the large game in both North and South America."[5] Thus while there is the pretense of speaking in general terms, it is American Indians, still living today, who are the object of the accusation. And how are Europeans treated? Some scholars find them as wasteful, but generally they are excused from blame. Kazimierz Kowalski writes, "Man was present at that time in Europe, but there is almost no evidence for his role in extermination of animals; at such a primitive stage of culture he was probably not a real danger for animals as large as an elephant or rhinoceros. The disappearance of these large mammals at the end of the last interglacial is the last episode in the great extinction of European fauna caused by Quaternary climatic changes in conjunction with the geographical configuration of the continent."[6] In other words, Europeans can blame the weather; Indians cannot even suggest it as a cause.

Robert Ebisch of *The Seattle Times* published a story on the overkill thesis in May 1990 and secured comments from a number of people. Stephen Pyne, a professor of American Studies at Arizona State University, expanded the thesis enthusiastically. Without realizing the background of the issue, Pyne was quoted as saying: "It wasn't just the spears. They also had torches, and fire is a pretty powerful tool. The largest source of ozone and nitrous oxide in the world today is coming from the savannas in Africa that are burned annually, for hunting, slash and burn agriculture, etc.,"[7] thereby unwittingly bringing the debate full circle to Sauer's discredited thesis.

Jared Diamond has written one article on the so-called big-game hunters and mentions the overkill thesis favorably in another article which was generally slanted in a proexploitation-of-environments alignment. He is enthusiastically in favor of blaming the big-game hunters, as they have come to be called, and likens them to the people who exterminated the giant moa in Australia but not to the British colonists who eliminated the Tasmanians. I will save his comments for the in-depth discussion of Martin's thesis, since he offers additional arguments on its behalf.

So what are we really talking about? Let us look first at Paul Martin's original compilation of papers discussing the extinction of animals, then at comments made by fellow writers in the book. Having established the basic framework, we will then look at some of Martin's later elaborations on the idea. And we will raise some questions about the viability of the idea in rebuttal and critique.

Pleistocene Extinctions is roughly divided into three parts: introductory statements, a set of essays under the rubric of "Facets of the Problem," and a set of essays under the classification of "Regional Aspects and Case Histories." Before we examine Martin's argument, however, it is a good idea to list a representative sample of the species that do become extinct at some point in the recent past (perhaps 50,000 years ago) so readers will have some idea of the scope of the thesis and what it tries to cover. There is a process of selection that occurs when Martin develops his thesis that should not be missed. The species who became extinct that would most easily fit into his framework were:

Artodus—giant short-faced bears
Bootherium—extinct bovid
Boreostracon—glyptodon
Bos—yak
Brachyostracon—glyptodon
Camelops—extinct camels
Castorides—giant beaver
Cervalces—extinct moose
Chlamytherium—giant armadillos
Cuvieronius—extinct mastodons
Dinobastis—saber-toothed cat
Equus—horses
Eremotherium—giant ground sloth
Euceratherium—shrub oxen
Glyptodon—an armadillolike creature
Glyptotherium—glyptodons
Hayoceros—extinct pronghorn
Hydrocherus—capybaras sloths
Mammut—American mastodons
Mammuthus—mammoths
Megalonyx—ground sloths
Mylohyus—woodland peccaries
Neochoerus—extinct capybara
Nothrotherium—small ground sloths
Paramylodon—ground sloths
Platyerabos extinct bovid
Platygonus—extinct peccary

Preptoceras—shrub oxen
Sangamona—caribou
Siaga—Asian antelope
Smilodon—saber-toothed cat
Stockoceros—extinct pronghorn
Symbos—woodland musk ox
Tanupolama—extinct llamas
Tapirus—tapirs
Tetrameryx—extinct pronghorns
Tremarctos—spectacled bears

This list is not complete, but it is certainly representative of the loss of species. There are still a good variety of grazing herbivores who became extinct to be considered, and the popular carnivores, including the dire wolves, also suffered extinction during this period of time but have not been mentioned.

Martin more or less stacks the deck in his essay because his list of smaller species which became extinct is drawn primarily from the Pliocene era, not the Pleistocene, which makes it appear that megafauna alone suddenly vanished and that smaller fauna had already passed from the scene at an earlier date. No mention is made of avian species, who could not have been exterminated by hunters under any circumstances. So the question of extinction in the late Pleistocene has already been restricted to those species, and those species only, that Martin believes will make his case, leaving to others the necessity of accounting for the demise of the remaining species.

The basic problem is posed as follows: Martin's thesis is that thirty-one species of herbivores weighing over fifty kilograms became extinct *sometime* during the late Pleistocene. He rejects climate change as a possible source of their demise, pointing out that a lesser number of megafauna species became extinct on other continents. But he does not say how many species of megafauna actually inhabited other continents and also became extinct, thereby magnifying the scope of disaster in North America. If, for example, Europe had a small number of megafauna originally, percentage-wise the depletion might have been as severe or worse.

As herbivore megafauna became extinct, so also did the megacarnivores that once fed upon them. Martin avoids the question of how carnivores became extinct by confining his inquiry to the loss of herbivores. Presumably with prey declining precipitously, the megacarnivores first reduced their population and then suffered radical and permanent decline for lack of food supply. Thus carnivores do not receive blame for the extinction of megaherbivores even though they would probably have been the prime users of the grazing animals.

Martin's case becomes one of suggesting, without clearly stating, that the only remaining possible factor in the extinction of megaherbivores would have been Paleo-Indian hunters, who he believes came into North America at this time and ruthlessly slaughtered the grazing megaherbivores. The slaughter was made possible by the fact that these creatures had never seen human beings and were therefore helpless to protect themselves. Inherent in this proposition is the belief that herbivores could not or did not recognize danger when it approached them, whether in the form of a saber-toothed cat or a few hunters waving spears and stone clubs.

Since most American anthropologists accepted the Neanderthal to Cro-Magnon evolution, the late entrance of man into North America was a given. Clovis-point locations, which incidentally are scattered all over the western United States on the surface as much as buried, and which by the common agreement of scholars date to around 12,000 years ago, then enabled Martin to argue that "the Indians did it" by linking a few sites which had bones of extinct megafauna and were also dated at that time.

The thesis is really applicable only to the herbivores, however, because almost every advocate of the idea cites those locations where mammoth bones are associated with evidence of human activity. From the list above we never hear about the giant rhinoceros, giant beaver, or giant armadillo, nor do the scholars refer to carnivore extinction except by indirection, assuming that the extinction of herbivores doomed meat-eating predators. Can we imagine hungry saber-toothed tigers and other carnivores unable to feed upon the smaller species of deer, moose, and bison when they discovered that the mega-animals had been destroyed?

When the Europeans came to North America the land was filled to overflowing with all manner of edible grazing game. The bison are conservatively estimated at a population of nearly 60 million creatures at the time of discovery. Since no species could evolve in 12,000 years, we must assume that the game animals we see today were here in their present form at the time when Martin suggests the Paleo-Indians were ruthlessly slaughtering the mammoth and mastodon.

So we have actually two questions. Why did the megacarnivores not pounce upon the smaller, weaker herbivores and maintain themselves in grand style? Why did the Paleo-Indian hunters not begin with smaller-sized animals, which would have been easier to kill, less dangerous to be around, and which themselves might be relegated to the fringes of the good grazing places by the larger and certainly more dangerous megaherbivores? Martin made a feeble effort to answer the second question by admitting that "we must beg the question of just how and why prehistoric man obliterated his prey. We may speculate but we cannot determine how moose, elk, and caribou managed to survive while horse, ground sloth, and mastodon did not."[8] He begged people not to ask him for specifics about the second question and was not even aware of the complexity of the first question.

Martin did, strangely, provide an opening for criticism of his idea by admitting: "What would upset the hypothesis of overkill would be clear-cut cases on the continent of many of the extinct animals surviving beyond the time of the big-game hunters, or clear-cut cases of massive unbalanced Pleistocene extinction anywhere before man."[9] I will provide evidence on this point in the next chapter, but it is sufficient here to note that Indian traditions and recent sites for some of the megaherbivores are rejected by most scholars as being too late. That is to say, very early and very late dates for mammoth and other creatures are cast out when citing evidence for the theory, apparently because of doctrinal considerations.

Martin is no exception to this scholarly rule. In some areas of Alaska are deposits, politely termed "muck," which are simply large piles of broken animal bones, trees, some volcanic ash, and gravels. They played hell with hydraulic mining because they would give off a terrible stench akin to ammonia when they were melted. Martin deals with this evidence in a peremptory way: "Stratigraphic chaos, apparently the result of intense solifluction, has thus far defeated attempts at direct radiocarbon dating of the abundant remains of extinct horse, bison and mammoth in Alaskan muck. ... The dates, on wood associated with the bones, appear much too young to represent the true age of the fauna and are thought to be intrusive."[10]

"Intrusive" is a favorite concept for scientists who find that the data do not conform to theory. It means that you argue that *after* the deposits you are examining have been laid down, trees, coins, artifacts, and bones which were later deposited on the surface begin to worm their way into the soil and burrow until they become jarring discrepancies in the strata you are considering. If the muck deposits are a twisted mixture of intertwined bones, trees, and gravels, how can the wood be intrusive? Thus Martin explains away evidence that would conflict with his theory by a spurious argument phrased in traditional scientific lingo.

Another source of megafauna bones which have a relationship to the big-game-hunter thesis once existed at Big Bone Lick, Kentucky. With respect to this evidence Martin writes: "In the case of Big Bone Lick, Kentucky, one of the richest late Pleistocene deposits in eastern North America, two samples of wood initially thought to be associated with bones of extinct species proved modern."[11] So what? Maybe they were modern; maybe these creatures weren't exterminated by hunters at all and survived until modern times. Fortunately, we have evidence of this survival recorded. George Gaylord Simpson discussed this site in an essay on fossils, and this site will be seen to have significance for our discussion:

> In 1762 John Bartram, the Philadelphia Quaker who supplied Linnaeus and other European naturalists with descriptions and specimens of American plants and animals, heard that a large tooth and a fragment of a tusk had been brought to Fort Pitt by some Indians. He requested his friend James Wright to make inquiries among the Indians concerning the place where these objects had been found. Through an interpreter, Wright secured an account of the site which came to be known as Big Bone Lick. According to Wright's informants, the lick contained five entire skeletons, the heads pointing toward a common center. The bones were of enormous size and were accompanied by tusks ten or twelve feet long. No such creatures as these had ever been seen alive by the Indians, but legend said that they had once been hunted through the forests by men of gigantic stature and that when the last of these men had died, God had destroyed their mighty prey in order to protect the present race of Indians."[12]

We cannot date the giant men and we cannot say when these giant animals were finally destroyed, but we know that the Indians did not have a hand in their demise. If the bones are fresh, then Martin's rejection of them as modern is simply silly. They might be modern.

Ashley Montagu and others published articles in the 1940s suggesting that mammoths were still alive when white men began to move west across the Ohio River and early colonists collected some stories suggestive of mastodons. It would seem logical that if these animals were killed thousands, or even hundreds, of years ago, their bones would have decayed and disappeared. How bones of extinct animals could lie on the ground for any period of time without disintegrating is an objection that cannot easily be dismissed.

Other essayists in *Pleistocene Extinctions* generally seem to be sympathetic to Martin's thesis, but they admit points so discordant with his ideas that we must note them. James Hester attempted to deal with the question of mammoth extermination and his analysis is worthy of note. "The staple food during the Llano period consisted of the mammoth; during the later periods it was the bison. All other animals are poorly represented. The size of the kills varies from site to site, and it is not always possible to determine if the animals present were killed at one time or on several occasions. In addition it is not always possible to determine if the animals present at the site were killed by man or some other agency."[13] There is, consequently, no clear and irrefutable evidence that Paleo-Indians ever killed more than one animal at a time, or that kills were numerous enough to even attract the attention of scholars.

Hester also pointed out that Folsom hunters generally chose small herds of bison to hunt, without discrimination as to age and gender. Clovis hunters would probably have followed the same general procedure. "In three separate kills at the Clovis site, the kills took place while the animals were knee deep in a pond, presumably drinking."[14] This tenuous form of hunting is a far cry from massive herd extermination. N. K. Vereshchagin, in his essay, pointed out that Paleo humans were capable of killing large animals but we know nothing of their weapons. "Stone instruments of the Lower Paleolithic, large hand points (ovates) and rough narrow-pointed stones, were unsuitable for killing and cutting up large animals. A sleepy cave bear could be killed with a huge wooden cudgel or boar spring ... but certainly not a mammoth or European Bison."[15] We can imagine Paleo-Indians swarming into caves to dispatch sleepy cave bears and perhaps even saber-toothed cats, but in our wildest dreams we cannot see how they could exterminate a species by this method of depredation.

So, then, how did Paleo-Siberians hunt mammoth? According to Vereshchagin, "usually, communities of hunters armed with spears hunted for solitary mammoth or small groups of females with young, as African Negroes now hunt elephants. The hunters pierced the abdominal wall of the mammoth and then pursued them, sometimes for many kilometers; but hunters could obtain only single specimens at a time. They hunted woolly rhinoceros in the same manner."[16] Surrounding a solitary mammoth, trying to pierce the abdominal wall with a spear to induce internal bleeding, and then chasing the creature until hemorrhage weakens or kills it would not seem to induce a massive destruction of species—certainly not cause the extinction of the thirty-one herbivore species that were lost somewhere in the Pleistocene.

Vereshchagin describes the effect of a hunting group traversing lands occupied by herds of grazing animals. He says that "after great but short-term migration of nomads, the population of large animals in steppe and forest-steppe did not disappear but relatively quickly returned to their former numbers. In the southern part of eastern Europe the abundance of steppe and forest animals was maintained through the Middle Ages."[17] So while hunting groups can temporarily reduce the numbers of a specific animal, what we know of hunting practices suggests that hunters cannot exterminate a single species, let alone a large number of species.

Vereshchagin also deals with the massive numbers of animals in the Siberian frozen wastes, but he does so directly and not by dismissing them. He notes that "the accumulations of mammoth bones and carcasses of mammoth, rhinoceros, and bison found in frozen ground in Indigirka, Kolyma, and Novosibirsk islands bear no trace of hunting or activity of primitive man. Here large herbivorous animals perished and became extinct because of climatic and geomorphic changes, especially changes in the regime of winter snow and increase in depth of snow cover."[18] If we follow the Martin scenario, these Paleo-Indian hunters could have and should have exterminated the mammoths, the woolly rhinoceros, and the ground sloth, among others, in Siberia without having to travel across the Bering Strait but did not do so. *On their arrival in Alaska, we are asked to believe, an incredible blood lust appeared and Paleo-Indians changed their hunting techniques in order to slaughter herds of mammoth, bison, armadillos, for example, with great efficiency.*

Arthur Jelinek, Martin's colleague at the University of Arizona, suggested that the Paleo hunters did have great efficiency and that this "is attested by such sites as Solutre in east-central France, where a late-Perigordian level is estimated to contain the remains of over 100,000 horses."[19] Other scholars give much lower estimates of the number killed, but generally this European site is believed to provide evidence of the blood lust of Paleo people at this time. But to cite a location in Europe as evidence of American Indian blood lust is not making a good argument. It is convoluted and bordering on irrational academic racism. But a much larger question arises: How in the world could primitive hunters, or anyone else, gather together 100,000 horses for any purpose? And would not these Paleo people have littered Europe with carcasses even before they traveled to Siberia? Scholars credit these ancients with feats that modern people would be hard pressed to duplicate.

Jelinek, although very supportive of Martin, noted a glaring lack of evidence for the theory, a defect that Diamond, Ardrey, and Bakker apparently did not catch. Jelinek admitted: "Throughout the New World one major puzzle exists with regard to linking man with the extinctions. *This is the absence of direct evidence of human activity associated with the remains of extinct animals. In fact we have kill sites with implements in association with partially articulated skeletons for only one of the many genera that disappeared in western North America and Mexico—Mammuthus.*"[20] (Emphasis added.) So why is this "theory" an explanation for the loss of many species of animals, including but not strictly limited to megafauna, in the Pleistocene? Is it simply because Martin is a tenured academic in good standing with his peers? Does *anything* that a member of the scientific establishment says receive uncritical acceptance by his peers as scientific even if it doesn't make any sense at all? If there are so few sites linking ancient man with extinct animals, what are we talking about?

In 1990, Paul Martin published yet another article on his energetic and mischievous hunters entitled "Who or What Destroyed Our Mammoths?" It had been twenty-three years since his first essay, and, while he had published a variety of papers in the interim advocating his ideas, in this essay he waxed poetic. Instead of filling in the immense gaps in evidence that characterized the first and subsequent articulations of the theory, Martin devised a scenario which he believed gave a better description of his theory. The latest version of the tale will unquestionably be cited as a serious elaboration and perhaps final proof of the overkill thesis by the many popular science writers, although Martin himself couches it in terms of his "best guess" and calls it a scientist's "bedtime story." We will look at portions of the most recent Martin scientific venture.

> Once upon a time 12,000 years ago, a small band of people physically very like us and speaking a common language trekked farther east than anyone had ever gone before. ...

> In autumn they followed their prey—mammoth, bison, and caribou—to wintering grounds. Helped in the hunt by opportunistic wolves or wolfish dogs, the hunters were expert at locating and tracking game, at killing it in excess of their needs, at butchering the carcass, and preserving the meat. ... At 60° below, they dressed in warm furs and slept in bearskin bags. ... By sharing food, there was always more than enough to eat and large animals were easily killed just for the fun of it, although wise elders spoke against this. ...

> Spreading southward, the unwitting explorers found that they were in a hunter's paradise, a Garden of Eden vastly rich in resources of the kind they were so expert in hunting. Some of the animals they discovered were new to their experience and some of these, three great ground sloths, were slow-moving and extraordinarily easy to dispatch. The children of the hunters could use them for target practice.[21]

I need not repeat any more of this dreadfully silly scientific scenario, but I can comment on these few points without fear of losing my bearings.

Hunters, in Europe, Asia, Siberia, and presumably even North America, moved south to north along river valleys. Northern

Siberia already had megafauna in profusion which never felt the spear or butchering tools, so we know that the scenario of extinction of so many species is unlikely at best. If these hunters were going on a large-scale program of extermination they could have done it in Siberia; they did not need to travel to Alaska to do it.

Martin's hunters move from autumnal regions to wintering grounds which have temperatures of 60° below zero. Now, Martin is from Arizona and consequently may have no firsthand knowledge of cold, so we can provide him with some helpful hints on how to improve his theory. At 60° below zero you are not out hunting mammoths, bison, and caribou. Everything is pretty much stationary and even a short exposure to these kinds of temperatures on bare flesh will give a pretty bad case of frostbite. Thus, to imagine hunters who have killed a mammoth deliberately going back out into this freezing cold to kill several more "just for the fun of it" is difficult at best when we consider that the method used to dispatch the beast is that of piercing the animal's abdominal cavity and chasing it until it bleeds to death. It is useful also to note that the coldest place on Earth today is Antarctica, which has temperatures of 70° below zero.

We are not told in this new Martin scenario what these animals are eating, but we can assume that there is some kind of grass in the area even at 60° below. But what about the helpful wolves and wolfish dogs? They are also predators and hence competitors of man, and it is unlikely that they are working in concert with these happy late Pleistocene hunters. More likely, if they are around at all, they are lurking in the hills waiting for humans to finish their hunt and hoping for leftovers. In the alternative, Martin would have to have these animals already domesticated, since without man they would have a terrible time scaring up a meal in that weather.

Why also do these hunters even bother to preserve food? At 60° below you have a reasonably good fast-freeze environment so that preserving food should be the least of your worries. Indeed, the problem would be finding enough fuel to keep a fire going to defrost supper. And if the hunters wastefully kill mammoths and never worry about having food, exactly what kind of food is being preserved?

We move southward with these hunters and discover new kinds of animals, and the kids use giant ground sloths as targets because they are so slow. Martin here shows an abysmal lack of knowledge of hunting societies. Children matured, or took on adult kinds of responsibilities reasonably early in life, especially in Indian tribes. Luther Standing Bear in his books relates that by the age of ten he had killed a buffalo, gone on a war party, and been inducted into the Fox Society, an honorary group that performed a charitable function in Sioux society.

If boys were performing some adult functions by age ten, then the children who were knocking off the ground sloths must have been between five and nine years old. Since some of these ground sloths were the size of modern adult giraffes, we can conclude that these hunters' children were among the more precocious and ambitious youngsters any society has ever raised.

Martin has his hunters moving two hundred miles in a generation and he figures that the hunters reached Tierra del Fuego within five hundred years. The distance between Alaska and Tierra del Fuego must be in the neighborhood of eight thousand miles, so figuring two hundred miles per generation, we must have had around forty generations of people within a five-hundred-year period, or a generation every twelve and a half years. So not only were these people expert hunters, they were something to behold in those bearskin bags.

Martin's scenario is simply preposterous. There is no point at which it makes any sense at all. No convincing evidence is cited except that he believes something happened at 10,000 or 12,000 B.C. that eliminated the megafauna; he looks only at the evidence he has amassed in support of his theory and then insists that the Indians did it. Bakker, Diamond, Ardrey, Ebisch, and a good many others cannot have read the original essay or the updated scenario of helpful wolves and gigantic target sloths. If they had read it, surely they would not so casually endorse the overkill hypothesis.

In the next chapter, I will examine other evidence relevant to the overkill theory. Obviously many of the mega-animals were destroyed or eliminated. Eastern Indians say that when the white giants went away, the Great Spirit took pity on them and destroyed the giant animals. In view of Martin's latest scenario, the Indian version sounds a lot more rational, and we will see that indeed it is.

6

The Corpora Delicti
and Other Matters

JARED DIAMOND, AS PREVIOUSLY NOTED, became a big booster of the overkill thesis, entitling one of his articles "The American Blitzkrieg: A Mammoth Undertaking." Spinning out his own scenario, he followed Martin in suggesting that "as the mammals were killed off in one area, the hunters and their offspring fanned out into new areas that still had an abundance of mammals, and kept exterminating the mammal populations at the front of their advance,"[1] a feat clearly contrary to the practice and capability of all observed hunting groups. In fact, even using the idea of "extermination" implies that the hunters deliberately swept the plains, valleys, or river bottoms bare of animals, and that condition would be impossible to guarantee.

One of Diamond's comments in this article is symptomatic of the manner in which scientists have tried to indict Paleo-Indians for this massive extinction: "Juries have convicted murderers on less compelling circumstantial evidence."[2] *There have been times, particularly in the American South at the end of the last century, when a simple accusation against an African American was enough to ensure a conviction. What scholars have done, therefore, is something akin to a southern lynching since there is little evidence that Paleo-Indian hunters did anything more than occasionally catch a mammoth at a waterhole.* Diamond's sympathy, however, lies with the archaeologists who continue to indict the Paleo-Indians. Unwittingly, Diamond betrays his own case when he writes:

Archaeologists trying to document the slaughter today are searching for needles in a fossil haystack: a few years' worth of butchered mammoth bones among the bones of all the mammoths that died naturally over hundreds of thousands of years. It's no wonder—given all the forces arrayed against paleontologists—that so few mammoth carcasses with Clovis points among the ribs have been found.[3]

On reading this passage one is tempted to pity the poor scholar who has to sort through an immense pile of mammoth bones in order to find one or two that have the carving marks of man. If these Paleo-Indian hunters had been so efficiently destructive, wouldn't the problem be reversed? Shouldn't archaeologists be looking at hundreds of thousands of bones with butchering marks, trying to find an unmutilated femur? Apart from an occasional bad review, what are the "forces" arrayed against paleontologists? Common sense, perhaps?

Here Fred Warshofsky can possibly assist Diamond. There were no bones to sort through. Writing his own version of the big game hunters thesis, he says that "so sudden and massive was the assault that the typical piles of animal fossil bones and associated human stone tools and other cultural materials that mark the kill sites of prehistoric man in Europe and Asia were rarely found in North America."[4] Now, is it possible to launch a massive assault on a species and leave no trace of evidence? Logic would dictate that the more furious the effort, and the shorter the time allowed to complete the task, the more evidence would exist documenting the crime. Could Paleo-Indians have been that tidy?

In even the most prejudiced murder trial there is one essential element: there has to have been a killing. Fancy legal terminology generally requires a body—the corpus delictus as the TV detective shows are fond of telling us. It would seem reasonable, if one was to promulgate a theory of blitzkrieg slaughter as have Martin and Diamond, to identify where the bodies are buried and then take the reader on a gut-wrenching tour through a graveyard of waste and butchery. We are deprived of this vicarious thrill because the evidence of the destruction of the megafauna suggests a scenario well outside the orthodox interpretation of benign natural processes. Therefore mere mention of the reality of the situation is anathema to most scholars. So let us see what the actual situation is.

The first explorers of the northern shores of Siberia and its offshore northern islands and of the interior of Alaska, and some of its northern islands, were stunned to discover an astronomical number of bones of prehistoric animals piled indiscriminately in hills and buried in the ground. The graveyards of these animals were classified as "antediluvian" (prior to Noah's flood) by the majority of scientists and laypeople alike who still believed the stories of the Old Testament. Near these graveyards, incidentally, but located in riverbanks on the northern shore of Siberia, are found the famous Siberian mammoths whose flesh was supposedly edible when thawed.

Reading an extensive set of quotations is always tedious to readers but I hope you will bear with me in this chapter because it is only in the repetition of the reports of the discoveries of these areas that the entire picture of the demise of the mammoths and other creatures really becomes clear. These Siberian remains are not the thousands of mammoth bones which Jared Diamond thinks are searched frantically by archaeologists seeking signs of human butchering. It is doubtful that any archaeologists or paleontologists have made extensive studies of the skeletons in these locations or we would certainly have a far different view of megafauna extinction than is presently acceptable to orthodox scholars.

Russian expeditions to Siberia and the northern islands of the Arctic Ocean began in the latter half of the eighteenth century, and with the discovery of these large mounds of animal bones, most prominently the tusks of mammoths and other herbivores, franchises were given to enterprising people who could harvest the ivory for the world market. Liakoff seems to have been the first important ivory trader and explorer in the late eighteenth century.

After his death the Russian government gave a monopoly to a businessman in Yakutsk who sent his agent, Sannikoff, to explore the islands and locate additional sources of ivory. Sannikoff's discoveries of more islands and his reports on the animal remains found there are the best firsthand accounts of the Siberian animal graveyards.

Hedenstrom explored the area in 1809 and reported back on the richness of the ivory tusks. Sannikoff discovered the island of Kotelnoi, which is apparently the richest single location, in 1811. Finally, the czar decided to send an official expedition and from 1820 to 1823, Admiral Ferdinand Wrangell, then a young naval lieutenant, did a reasonably complete survey of the area. Since these expeditions and explorations were inspired by commercial interests and not scientific curiosity, the reports are entirely objective with no ideological or doctrinal bias to slant the interpretation of the finds.

Around the turn of the century (1905), interest in the Siberian islands seems to have increased, whether as a result of the few Christian fundamentalists who were not reconciled to evolution frantically searching for tangible proof of Noah's flood, or as part of the leisure activities of the English gentlemen of the time, we can't be sure. The definitive article on the Siberian prehistoric animal remains was written by the Reverend D. Gath Whitley and published by the Philosophical Society of Great Britain under the title "The Ivory Islands in the Arctic Ocean." It drew on older sources, primarily reports of expeditions of the ivory traders, and captured the spectacular nature of the discoveries well.

Liakoff discovered, on an island that now bears his name, rather substantial cliffs composed primarily of frozen sand and hundreds of elephant tusks. Later, when the Russian government sent a surveyor, Chwoinoff, to the island, he reported that, with the exception of some high mountains, the island seemed to be composed of ice and sand and bones and tusks of elephants (or mammoths) which were simply cemented together by the cold. Whitley reported:

Sannikoff explored Kotelnoi, and found that this large island was full of the bones and teeth of elephants, rhinoceroses, and musk-oxen. Having explored the coasts, Sannikoff determined, as there was nothing but barrenness along the shore, to cross the island. He drove in reindeer sledges up the Czarina River, over the hills, and down the Sannikoff River, and completed the circuit of the island. All over the hills in the interior of the island Sannikoff found the bones and tusks of elephants, rhinoceroses, buffaloes, and horses in such vast numbers, that he concluded that these animals must have lived on the island in enormous herds, when the climate was milder.[5]

Hedenstrom explored Liakoff's island in 1809 and discovered that "... the quantity of fossil ivory ... was so enormous, that, although the ivory diggers had been engaged in collecting ivory from it for forty years, the supply seemed to be quite undiminished. On an expanse of sand little more than half a mile in extent, Hedenstrom saw ten tusks of mammoths sticking up, and as the ivory hunters had left these tusks because there were still other places where the remains of mammoths were still more abundant, the enormous quantities of elephants' tusks and bones in the island may be imagined."[6] Indeed, a number of explorers reported that after each ocean storm the beaches were littered with bones and tusks, which had been lying on the sea bottom and brought to shore by wave action.

The elephant or mammoth bones and tusks were the most spectacular finds primarily because they were so plentiful and consequently they attracted public attention the most. The islands contained an incredible mixture of bones of many extinct and some living species of mammals. Mixed with the animal bones were trees in all kinds of conditions. Whitley quoted some of the Russian explorers as reporting "it is only in the lower strata of the New Siberian wood-hills that the trunks have that position which they would assume in swimming or sinking undisturbed. On the summit of the hills they lie flung upon another in the wildest disorder, forced upright in spite of gravitation, and with their tops

broken off or crushed, as if they had been thrown with great violence from the south on a bank, and there heaped up."[7]

A few conclusions can be drawn from the reports of the Russian
ivory traders. First, it appeared that several reasonably large islands were built primarily of animal bones, heaped in massive hills
and held together by frozen sand. To indicate the scope of the debris, we should note that all of these islands are found on modern
maps of the area, indicating that we are not talking about little
tracts of land of limited area. Second, the sea floor north of Siberia
and surrounding the islands was covered with so many additional
bones that it was worthwhile for the ivory traders to check the
beaches after every storm to gather up tusks and other bones.

Third, and very important for estimating the scope of the disaster, the ivory was of outstanding quality, so much so that the area
provided most of the world's ivory for over a century. Estimates of
the number of tusks taken from the islands between the 1770s and
the 1900s range in the neighborhood of 100,000 pairs. Whitley
noted that Sannikoff himself had brought away 10,000 pounds of
fossil ivory from New Siberia Island alone in 1809.[8] In reality, however, only about a quarter of the ivory was of commercial grade, so
the true figure must approach half a million pairs of tusks.

Fourth, an amazing variety of animals, many extinct, were mixed
with the mammoth and rhinoceros bones, although these two
animals have become symbolic of the whole menagerie. Fifth,
trees, plants, and other floral materials were indiscriminately
mixed with the animal remains, sometimes leading the Russians
to suppose that the islands represented a sunken isthmus or broad
stretch of land where these animals and the companion plants
lived in a warmer climate. The chaotic nature of stratification of
the remains soon abused that notion.

Finally, it is important to note that *none* of the bones of any of
the species had carving or butchering marks made by human beings. N. K. Vereshchagin wrote: "The accumulations of mammoth
bones and carcasses of mammoth, rhinoceros, and bison found
in frozen ground in Indigirka, Kolyma, and Novosibirsk lands bear

no trace of hunting or activity of primitive man. Here large herbivorous animals perished and became extinct because of climatic and geomorphic changes, especially changes in the regime of winter snow and increase in depth of snow cover."[9] The "climatic and geomorphic changes" must have been very sudden indeed and exceedingly violent, considering the fact that these bones are always described as "heaps" of material deposited as if they had been thrown into a pile by an exceedingly strong force.

The testimony regarding the richness of the animal remains in the Arctic north of the continental masses is not restricted to Russian sources. Stephen Taber, writing in his report "Perennially Frozen Ground in Alaska: Its Origins and History," had this to say about the Siberian islands:

> Pfizenmayer [citation omitted] states that in the New Siberia island collectors have "found inexhaustible supplies of mammoth bones and tusks as well as bones and horns of rhinoceros and other diluvial mammals"; and Dr. Bunge, during expeditions in the summers of 1882–1884, "gathered almost two thousand five hundred first class mammoth tusks on the new Siberian islands of Lyakhov, Kotelnyi, and Fadeyev," although many collectors had previously obtained ivory from the islands since their discovery in 1770 by Lyakhov.[10]

It would seem obvious to anyone seriously pursuing the question of the demise of the mammoth and the other megaherbivores that a good place to locate the bodies to determine the cause of their demise would be the islands north of the Siberian peninsula. Yet we hear not a word about them in scientific articles and books concerning the overkill hypothesis.

When we inquire if the Alaskan area has similar deposits, we learn that the situation is the same. Early gold miners in Alaska discovered that in many cases they had to strip off a strange deposit popularly called "muck" in order to get to the gold-bearing gravels. The muck was simply a frozen conglomerate of trees and plants, sand and gravels, some volcanic ash, and thousands if not millions of bits of broken bones representing a wide variety of late Pleistocene and modern animals and plants.

Two scholars describe the scenes of destruction and chaos which the muck represents. Frank Hibben, in an article surveying the evidence of early man in Alaska, said that while the formation of muck was not clear, "… there is ample evidence that at least portions of this material were deposited under catastrophic conditions. Mammal remains are for the most part dismembered and disarticulated, even though some fragments yet retain in this frozen state, portions of ligaments, skin, hair, and flesh. Twisted and torn trees are piled in splintered masses concentrated in what must be regarded as ephemeral canyons or arroyo cuts."[11]

Stephen Taber's report echoes the same conditions. He says: "Fossil bones are astonishingly abundant in frozen ground of Alaska, but articulated bones are scarce, and complete skeletons, except for rodents that died in their burrows, are almost unknown."[12] Many laypeople will be confused by this technical language and fail to grasp what Taber is saying, allowing him to imply a benign orthodox interpretation when the situation requires that a clearer picture be drawn.

When a scholar says "articulation" of bones he means an arrangement of bones that a person observing them would identify as a complete skeleton and from which an experienced observer could identify the species. To say that articulated bones are scarce, then, means that the bones are scattered and mixed so badly that expert examination is needed to identify even the bone itself, let alone the species from which it comes. Remember this problem of articulation, for we shall meet it again in the next chapter. Taber concludes with the observation that "the dispersal of the bones is as striking as their abundance and indicates general destruction of soft parts prior to burial."[13] In other words, Alaskan muck is a gigantic pile of bones representing a bewildering number of species, a good number of them the megafauna I have been discussing.

We find the missing megafauna of the late Pleistocene in the Siberian islands, in the islands north of Alaska, and in the muck in the Alaskan interior. Obviously we have here victims of an immense catastrophe which swept continents and left the debris in the far northern latitudes piled in jumbled masses that now form decent-sized islands. Most anthropologists and archaeologists avoid discussing these deposits because the orthodox uniformitarian interpretation of the natural processes precludes sudden unpredictable actions.

Paul Martin, in private correspondence with me in June 1993, stated flatly that the mammoths could not have been destroyed by any such force or event.[14] The sole basis he gave for that conclusion was radiocarbon dating of mammoth remains in the Siberian and Alaskan muck. I will have more to say about the reliability of radiocarbon dating below, but if we were to accept his argument, then we would have to create a scenario where Paleo-Indians kill all these animals without leaving a trace of a spear point or hatchet blade, drag the carcasses out to sea some 150 miles north of Alaska, and dispose of the evidence of their misdeeds. Here, friendly wolves would not be much help.

Although Martin maintains that his thesis explains the disappearance of the megafauna, his argument really centers on the loss of three species: mammoths, mastodons, and ground sloths, with an occasional reference to horses and camels that makes it appear as if the important species have been covered. But overkill avoids asking about the possibly half-million mammoth skeletons lying frozen in the Arctic regions because that would completely negate the theory.

What about the carnivores who disappear during this same period of time? Martin suggests that with the drastic reduction of herbivores, the carnivores also vanished, presumably of starvation, certainly not through voluntary reduction of the species until extinction. Again we should ask the question: Where are the bodies? A good many skeletons in Siberia and Alaska are carnivores, so that question is partially answered. But the most famous location of some carnivore species of the postglacial period is in the La Brea tar pits in Los Angeles, California. This location probably has the most numerous collection of saber-toothed cats and dire wolves of any place on Earth. So we should look at the La Brea carnivore skeletons to see if we can find any evidence of Paleo-Indian butchery or widespread malnutrition.

The story of the La Brea tar pits is more familiar to many schoolchildren than Bible stories. Long, long ago in southern California there were vast pools of asphalt seeping from the ground. More shallow pools of water formed on top of the asphalt creating a hazardous situation. When deer, bison, antelope, and other herbivores came to drink in the pools of water, they quickly discovered that the ground was not firm; it was 10w40-oil quicksand. They became trapped in the asphalt, which began to pull them down like quicksand, and they cried out in fear. Bleatings of distress attracted the neighborhood carnivores, who were not very smart either. They ran across the unstable surface to eat the helpless herbivores only to find themselves also trapped in the asphalt. Prey and predator slowly sank out of sight to be finally resurrected by curious archaeologists and paleontologists around 1900.

This scenario is firmly entrenched in our minds and we cannot rid ourselves of it because it seems so commonplace and sensible and because we have been told, since we were children, that it is one of the things that science knows as a certainty. We never asked why herbivores would not smell the asphalt and avoid an obviously unpalatable water source. We also failed to credit cautious predators with enough common sense to learn, over the generations, that La Brea was not as inviting as it might seem. These details are questions which, as a matter of common sense, we should have asked. But we didn't. In fact, it was not until I read some of the original articles on La Brea that I realized that the traditional explanation was wholly mythical and made no sense at all.

The truth is shocking. La Brea *never* had pools of oil standing beneath pools of water which attracted unfortunate beasts. "Stoner [citation omitted] concluded from the physical dimensions of the University of California pits 2050 and 2053 that neither would have been sufficiently stable to persist as continuously operative, open asphaltic pools during the long interval of alluviation accompanying the development of the Santa Monica Plain. A similar conclusion is warranted for the Museum excavations."[15] This analysis is taken from an excellent essay by Geoffrey Woodard and Leslie Marcus published in the *Journal of Paleontology* in January 1973. And they state unequivocally:

> In no reported cases were bones recovered from open, liquid asphalt lakes. Bones were entombed in hard, asphalt impregnated sand or clay. Where encountered, asphalt was confined to small unfossiliferous pockets within the tar soaked sediment. The fossil remains were frequently admixed with gravel lenses, cobbles, and pebble clasts of fluvitile origin. Freshwater limestone lenses, mollusks, and hardened asphaltum deposits were interbedded with bone bearing sediment in several of the pits. No bones were found in association with the limestone strata, and the bones count was diminished when hardened asphaltum was encountered.[16]

Instead of the popular scenario, then, what more likely happened is that a very large mass of bones and vegetal material, the result, perhaps, of a natural catastrophe, was covered with sand and clay and some time later, probably a very short time later, some kind of asphaltic or bituminous material in vast amounts was dumped on this location. The asphalt tar then seeped into the sand and clay and in depressions made small pools which hardened into asphalt deposits with the passage of time.

Most probably, very few animals were ever trapped in the asphalt waiting to be consumed by a predator, and we can say almost certainly that no megafauna was consumed by a saber-toothed cat or dire wolf who then found themselves sinking into the tar while chasing a trapped victim. The number of carnivores in comparison with herbivores found at La Brea shows an overwhelming number of carnivores. To follow the orthodox rendering of the story we should minimally have one herbivore for each carnivore so that the scenario—frightened cries of a trapped deer invite a minimum of one carnivore for supper—could be maintained.

Folks, we have been had by overly poetic science writers. But if there were no pools of asphalt, as we have all been taught, how about the pits that have been excavated by scientists? Doesn't the presence of the tar pits provide evidence of the traditional scenario?

Here is the irony of the situation. According to Woodard and Marcus:

> In the majority of publications concerning Rancho La Brea, the term "pit" has been used to imply the initial presence of an open, liquid asphalt filled crater or pool which became the primary focus for the entrapment and preservation of vertebrate animals. Actually the pits were excavations and as such the terms are synonymous.[17]

Now isn't that a pretty pickle? We call them the La Brea tar pits *only* because the *shape* of excavated holes is in the form of a pit. So it appears that several generations of popular science writers did not go to original sources but simply assumed that the pits were natural traps for animals and devised a scenario of unsuspecting prey and predators mired in a sticky petroleum pit from which they were unable to extricate themselves.

My research assistant called the Page Museum to get an updated census of carnivores found at La Brea. She learned that no accurate account exists but that three million different items, ranging from bones to plants and insects, have been recovered from the pits. John Merriam, an early observer of the La Brea tar pits, wrote in 1911: "While it is not safe to make any attempt at an estimate of the quantity of material represented in the whole area of the Rancho La Brea Beds, it is within the limits of probability to say that there are still some thousands of individuals of *Smilodon californicus* and *Canis indianesis* entombed in the deposits of this region. Of other forms the number is probably very large, though somewhat less than in the case of the Carnivora."[18]

That estimate was made over eighty years ago, indicating that the pits contained a substantial number of carnivores in 1910, out of proportion to the other fauna. The Page Museum is badly funded and grossly understaffed, so maintaining an accurate count of the skeletons retrieved is impossible. Nevertheless, in the more than eighty years since Merriam's article, there should have been some progress made in determining how many megafauna are represented in this formation, or whether the ratio of prey and predator has changed significantly.

The people at the Page Museum did refer us to Chester Stock's work, *Rancho La Brea: A Record of Pleistocene Life in California,* as the only possible place where a census of animals might be found. Stock obscured the question rather than answering it. He wrote: "A census of the Pleistocene mammals represented in the collection of the Page Museum reveals a total of more than 10,000 individuals."[19] But no charts were given, listing the comparative number of individuals for each species in the pits.

Looking through Stock's listing we find something amazing. Writing of *Smilodon fatalis,* the saber-toothed cat, Stock says that "this form ranks next to the dire wolf in number of individuals found in the asphalt seeps and greatly outnumbers all other types of cats. G. J. Miller [citation omitted] documented 2,100 *Smilodon* individuals from the Rancho La Brea collections at the Page Museum, based on crania and cranial fragments."[20] Now, if the dire wolves have a greater representation than saber-toothed cats, and the cats number 2,100 individuals, then dire wolves must run around 3,000 individuals, meaning that at least *half* of the Pleistocene fauna found at La Brea were predators. What then happens to our traditional scenario where the saber-toothed cat or dire wolf spots a hapless deer and races out to catch it, then becomes mired himself? There would have had to have been a virtual line of predators waiting to attack the solitary deer to account for these ratios of predator to prey at La Brea.

At La Brea, then, we are able to locate two of the megapredators who were supposed to have died out as a result of the mega-herbivore extinction—the saber-toothed cat and the dire wolf. They are the most common species in the assemblage and are unceremoniously mixed with masses of gravel, vegetable material, and asphaltic-soaked sands. How they died, we can only surmise. It hardly seems likely that the skeletons represent hundreds of years of time, since the bones would deteriorate quickly after the death of the individual. Whatever the cause of their demise, it is certainly a far cry from any hypothesis involving big-game hunters.

Martin's original essay in *Pleistocene Extinctions* admitted that his hypothesis would fall if there were "clear-cut cases on the continent of many of the extinct animals surviving beyond the time of the big-game hunters, or clear-cut cases of massive unbalanced Pleistocene extinction anywhere before man."[21] Many scholars, objecting to the evidence of mammoths and La Brea carnivores presented above, would offer as their rejoinder the contention that the Siberian and Alaskan deposits must be dated around 38,000 years ago and therefore represent an earlier but by no means fatal event which extinguished *some* of the Pleistocene megafauna. Such an argument, of course, falls into Martin's second qualification and voids his hypothesis. Let us turn, then, to the first proposition: to show that many extinct animals survived beyond the time of the mythical big-game hunters.

What about human testimony regarding megafauna? Here we have a very controversial subject because scientific doctrine suggests that the megafauna were destroyed so long ago that no human society could remember them. Not even monster or dragon stories of Western society are allowed standing as a possible memory of the megafauna. Most archaeologists and anthropologists automatically dismiss the traditions of non-Western people concerning the natural world and their experiences in it as superstitions unworthy of their attention. Consequently, it is very difficult to put forward oral tradition evidence that might provide the needed identification.

The Delaware Indian story of the Big Bone Lick skeletal remains in Kentucky cited in the preceding chapter suggests that there may be some recent Indian memories of these creatures. The Indians were collecting bones for European scientists so there should be no doubt as to their authenticity. Accepting the story, however, means that there were mammoths or mastodons still living in the eastern United States at the time the Pilgrims landed or when the Jamestown colony was being established—because the bones could not have lain on the ground for thousands of years without suffering complete decay and dissolution.

So where do we go to obtain convincing evidence that some Indian nations did have knowledge of some of the megafauna species? The Sioux people have a star constellation that is called the "Hairy Elephant" and, considering that the mammoth trap near Hot Springs, South Dakota, which has produced a good number of skeletons, was in the heart of the Sioux lands, there is every reason to suppose that this constellation reflects the memories and life experiences of these people with the mammoths.

On the whole, however, it is exceedingly difficult to find traditions about these creatures because Indians have different ways of describing unusual animals which do not always correspond to the characteristics that scientists and scholars would consider relevant. Let us look at some of the early efforts to discover memories of extinct Pleistocene megafauna in the tribal traditions.

In 1934, W. D. Strong published an article in the *American Anthropologist*, suggesting that some tribes may have preserved memories of the large animals, and he specifically identified the mammoth, primarily because the Eskimos and Abnakis, both northern peoples, had stories which purported to explain the presence of mammoth bones in the ground where they lived. The Eskimos came to believe that the mammoth bones in their vicinity belonged to a creature who lived underground, since the bones were a visible part of the landscape. Strong also cited several old accounts of white men captured by Indians who were shown large bones and told that the animals to whom they belonged once lived in the area.

Thomas Jefferson included two notable references to these stories in his *Notes on the State of Virginia*. In one account a Delaware delegation was visiting the governor of Virginia. Asked about the bones at the Salt Lick on the Ohio, the Delaware chief promptly assumed a dignified bearing and began his speech with all the formality of an elder passing along revered tribal knowledge:

> That in ancient times a herd of these tremendous animals came to the big-bone licks, and began a universal destruction of the bear, deer, elks, buffaloes, and other animals: that the great Man above, looking down and seeing this, was so enraged that he seized his lightning, descended on the earth, seated himself on a neighboring mountain, on a rock of which his seat and the print of his feet are still to be seen, and hurled his bolts among them until the whole was slaughtered, except the big bull. ...[22]

The bull dodged lightning bolts until he was wounded and then fled west where the Delawares believed he was still living. The other account presented by Jefferson is nearly as mysterious:

> A Mr. Stanley, taken prisoner by the Indians near the mouth of the Tanissee, relates, that, after being transferred through several tribes, from one to another, he was at length carried over the mountains west of the Missouri to a river which runs westwardly: that these bones abounded there; that the natives described to him the animal to which they belonged as still existing in the northern parts of their country; from which description he judged it to be an elephant.[23]

It can be argued that Thomas Jefferson was simply a gullible gentleman of leisure inclined to believe tall tales because of his curiosity. However, the ring of veracity in these accounts is bolstered by the presence of the bones at Big Bone Lick. They could not have been unreasonably old or they would have disintegrated lying on the surface of the ground.

Strong reported favorably on research done by Frank Speck on folklore of the Penobscot Indians of Maine which seemed to refer to mammoths, and himself relied on a Naskapi tale which incorporated a large animal into what was basically a morality story. While these kinds of stories receive some attention from folklorists, as a rule they do not move the physical scientists, and with good reason: a reprocessed account, even if it appears to be that of an eyewitness, is suspect because a basic story line has been created. We never know if the creature is included in the narrative because it was seen or because it was needed for the dramatic story line.

Many of the eastern woodlands stories characterized the great creature as a large moose, bear, buffalo, or elk rather than a mammoth, and it is impossible to tell if the Indians were referring to megafauna or simply an oversized and impressive representative of a modern species. The common characteristic which these beasts seem to share, regardless of tribal origin, is that they are always said to have been carnivorous. The Sioux and Cheyenne, for example, describe their giant bison as carnivorous, and the insistence on this trait is a constant annoyance to people trying to make sense of the tradition.

Sometimes the Indian stories are directly to the point and leave no doubt about what the people experienced. A tradition of the Chitimacha mentioned by Strong illustrates these kinds of stories:

A long time ago a being with a long nose came out of the ocean and began to kill people. It would root up trees with its nose to get at persons who sought refuge in the branches, and people lived on scaffolds to get away from it. It made its home in a piece of woods near Charenton, and when guns were introduced the people went into this wood to kill the monster, but could not find it. When the elephant was seen it was thought to be the same creature ... .[24]

This testimony about the contemporary existence of the mammoth should not be lightly dismissed.

Curiously, William Berryman Scott, a prominent early scholar, cited a tradition of the British Columbian Indian tribes to the effect that they had to build lakehouses on stilts (comparable to those found in Swiss lakes) to protect themselves from the mammoths.[25] The mammoth, then, does not seem to be the helpless herbivore unaware of the danger of human hunters stalking him. Rather, he seems to be an animal to be feared by people because of his tendency to destroy their homes and disrupt their lives. Strong concluded that the many traditions of which he had knowledge demonstrated a historical association between people and the mammoth.

Ludwell Johnson published a comprehensive overview of this subject in *Scientific Monthly* in October 1952, trying to expand on the subject. Johnson was responding to an article by Hugo Glass of the previous year which dismissed the possibility of man-mastodon associations. Citing a 1950 list of twenty-seven known associations, Johnson rejected the casual dismissal of the evidence suggesting contact and penned one of the most memorable rebuttal statements in the history of science:

> Perhaps the most devastating technique used to discredit discoveries which run counter to the general climate of anthropological opinion is to imply that the find in question is a hoax. The venerable brotherhood of the politely raised eyebrow could with a single skeptical glance strike a live mammoth dead in his tracks.[26]

In general, Johnson supported the idea that men and mastodons were contemporaries and suggested that the time scale, instead of being some 12,000 years ago, might be as recent as 3,000 years.

Initial efforts to identify which animal the Indians might be remembering have produced some strange arguments. The mammoth and mastodon both were stiff-legged, and consequently the first Indian reports of these creatures were passed off as if the Indians had read accounts describing similar animals which existed in Europe around the time of Julius Caesar. Both the northeastern Indians and the Romans had apparently cut through trees to weaken them so that, when the mammoth or mastodon leaned against the tree to sleep, it would fall over, trapping the animal because it would have great difficulty getting to its feet again.

F. T. Siebert, Jr., writing on mammoths and men, included an account of a creature which, for a short period of time, scholars called the "stiff-legged bear." But from the description he gives, this animal really seems to be an actual mega-bear and not a mammoth. The account is from David Zeisberger's *History of the North American Indians,* and Zeisberger was in a situation where he would hear news about strange animals:

> There is likewise a kind of bear, much larger than the common bear, with much hair on the legs, but little on the bodies, which appear quite smooth. The Indians call it the king of bears, for they have found by experience that many bears willingly follow it. While all the bears are carnivorous ... this kind of bear is particularly voracious. Many instances are known where they have seized upon even defenseless Indian women and children. In more northerly regions, as, e.g., in the country of the Mingoes [Iroquois], these are more frequently found and they have killed many Indians.[27]

How this creature could be classified as a mammoth or mastodon escapes me. It is carnivorous, not herbivorous, although in defense of this identification is the insistence of Sioux and Cheyenne peoples that the large bison was carnivorous. If it is a megafauna, then we have at least secondhand evidence that it survived into modern times at some locations. Perhaps it was the description of the upper body devoid of hair that convinced some scholars that the Indians really meant a mammoth or mastodon. But the Indians knew wild animals considerably better than non-Indians and would have recognized a bear when they saw one.

Ashley Montagu reprinted part of an article originally published by Robert Koch in 1841 reciting an Osage Indian tradition which seemed to be an eyewitness observation of interspecies warfare involving mastodons. Koch had already misidentified a ground sloth as a mastodon and seemed to be an enthusiastic booster of that species.

A short footnote in the article indicated that Montagu had the subject of man and megafauna much on his mind and that the scholars of his youth were exceedingly liberal about entertaining ideas of Indian knowledge of the megafauna. "In several conversations with the writer," Montagu confessed, "Professor William Berryman Scott, the doyen of American paleontologists, has given it as his opinion that, had the first of the Spanish discoverers of America penetrated into the interior, it is quite possible that they might have met with the living mammoth."[28] He also quoted, but did not cite by name, another prominent scholar who believed that horses never did become extinct—the Indians simply did not use them as beasts of burden until they saw whites doing it.

This story does not deal with mammoths, however, but with some unidentified creatures that greatly abused the Indians. Since other tribal traditions always describe megafauna as the culprits we can suggest them as the villans, but we have no way of knowing exactly which animals might be involved. Humans, however, are strictly observers and not actors in this event.

> There was a time when the Indians paddled their canoes over the now extensive prairies of Missouri, and encamped or hunted on the bluffs. ... That at a certain period many large and monstrous animals came from the eastward, along and up the Mississippi and Missouri rivers; upon which the animals which had previously occupied the country became very angry, and at last so enraged and infuriated by reason of these intrusions, that the red man durst not venture out to hunt any more, and was consequently reduced to great distress. At this time a large number of these huge monsters assembled here, when a terrible battle ensued, in which many of both sides were killed, and the remainder resumed their march towards the setting sun.

We can but imagine the intensity of the conflict involved. Knowing that herbivores do not act collectively in a strategic manner, it would appear that two different kinds of predators fought over territory. They remained on the battleground or in the vicinity until the issue was decided as prides or packs of predators might do. It seems clear from the Indian account that the eastern predators soundly defeated the western animals and moved through their hunting grounds toward the west. But the story continues and provides a few more important details:

> Immediately after the battle, the Indians gathered together many of the slaughtered animals, and offered them on the spot as a burnt-sacrifice to the Great Spirit; the remainder were buried by the Great Spirit himself in the before mentioned Pomme de Terre, which from this time took the name of the Big Bone river. ...[29]

The Indian memory of their response to these animals is worth noting. They feared these creatures. They did not swagger around intent on killing in excess of their needs or seeking to exterminate the species "just for the fun of it."

In recent decades, sites where it is evident that Indians and mammoths coexisted and the Indian hunters had killed and eaten some of these animals have been accepted. These sites are so few that to accept them as evidence of an overkill situation is absurd. Most scholars have now accepted man-mammoth contact and have raised questions about other megafauna that were supposedly exterminated by the Paleo-Indians. These speculative articles are welcome because they extend the range of possible man-megafauna contact. By always citing mammoth sites, Martin and his friends have succeeded in making the loss of the mammoth the equivalent of the loss of all thirty-one herbivore species.

Jane C. Beck posed a question in an article in *Ethnohistory* in 1972, "The Giant Beaver: A Prehistoric Memory?" Beck cited several tribal traditions concerning this animal, primarily from the northeastern United States, citing, as did Frank Speck and W. D. Strong, traditions of the Micmac, Penobscot, Malecite, and Wabanaki. Generally, she demonstrated that even when the beaver was included in folklore narratives, there was a basis for arguing contemporaneous existence of man with this animal. These tribes live in the northeastern United States.

The giant beaver was a spectacular creature about the size of an adult black bear of today, and some specimens ranged between six and nine feet in length and weighed in the neighborhood of three hundred to five hundred pounds. What they must have been able to do with a tree would have been very impressive, although no evidence exists that they cut down trees as do modern beavers. In general, the giant beaver seems to have been concentrated in the Illinois-Indiana-Ohio region according to scientists, although some skeletons have been found in the northeastern part of the United States. I have not found any tribal traditions about this creature from the Midwest, however.

At the time Beck wrote, only one specimen of the giant beaver had been found west of the Mississippi River, and she speculated that he might have become lost. It is singularly unfortunate that Beck did not discover Ella E. Clark's studies on Indian legends because many tribes of the Pacific Northwest have stories concerning the giant beaver that indicate widespread occupancy of that area by this creature. We will meet these giant beavers in chapter 8 where they will play an important role in our story.

Beck did reproduce an account by Pliny Goddard that has an important theme found in many other stories about the megafauna and is worth discussing. This story takes place on the Great Slave Lake in Canada and was told to Goddard by the Beaver Indians:

> ... there was a large man who chiseled for a large beaver. He worked in vain for he could not kill it. He could not find its track anywhere nearby. He went out on the large frozen lake and saw the beaver walking along under the ice. He tapped on the ice and drove the beaver back into its house where he killed it. She had young ones in her and because of that the ice would not remain quiet. He cut the mother open, took out the young ones, and put them in the water. The ice then became quiet. ... They say both the man and the beaver were giants.[30]

The story lacks poetic embellishment and is told in a matter-of-fact manner that suggests that the feat of killing the giant beaver struck these Indians as unusual, so the narrative was remembered.

One final witness must tell a story. Ludwell Johnson cited an anomalous bit of information in his article on elephants to broaden the discussion of men and extinct animals. He included a conversation between a Northwest Coast Indian elder and Marian Smith, a scholar who had been doing linguistic research in that area:

> While giving incidental vocabulary, a Puyallup informant named one too many members of the local cat family: cougar, panther, mountain lion, and a fourth lion. The Salish term for the latter, he explained, was currently used for the maned lion seen in the zoos, etc., but had enormous teeth. In his boyhood, people had told of the ferocious attacks of this cat on children and adults, although actual encounters, he insisted, had ceased "long, long before."[31]

Marian Smith pointed out the obvious—it was a saber-toothed cat—and Johnson approved the identification.

Many Indian tribes regarded these large animals with great awe because of the obvious physical talents they possessed. So they were reluctant to give information on them, knowing that the white scholars quizzing them would regard such information as a superstition. A great many stories have been preserved, and were elders comfortable in discussing this topic with people openly we should have considerably more accounts of the megafauna of the late Pleistocene era. An initial list of animals that might be identified would be the giant deer, elk, moose and bison, the mammoth, mastodon, giant beaver, the saber-toothed cat, and most probably the condors.

The American Indian Science and Engineering Society has been sponsoring small gatherings of traditional people in recent years to encourage them to make certain that this knowledge is passed down. Already a number of tribes have provided more information on several of the species discussed above. Additional information will become available as the traditional knowledge conferences continue to be held.

Approximately half of the Indian stories relating to these creatures state that a Higher Spirit took pity on the Indians and destroyed these animals—or at least made them go away. Giant men, quite often with white skin, are involved in about 20 percent of the big-animal stories that I have located and, if we accept them as Martin's mythical big-game hunters, then the Indian accounts might in some cases support Martin's thesis. But these mega-killers would probably be Europeans—demonstrating that even 12,000 years ago the whites had little regard for the environment or other forms of life, and that is a message that conservatives would not like to hear.

Martin, as you will recall, could not solve the problem of why the big-game hunters killed only megafauna and had no effect on the smaller-sized herbivores. Nor did he even attempt to deal with this question: Why, deprived of megafauna prey, did the large carnivore predators simply vanish when millions of other game animals were present? Some Indian stories offer an intriguing solution to these problems, and in the next chapter we will see what they suggest.

7

Creatures Their Own Size

THE GREAT SCIENTIFIC FALLACY OF "OVERKILL" should have fallen of its own inconsistency decades ago, particularly in view of Arthur Jelinek's admission that we simply do not, as a rule, find ancient man or his artifacts associated with the fossils of extinct herbivores. There are, of course, some Folsom and Clovis sites which contain mammoth remains, but it is unclear whether the human artifacts that are found there belong with these animal parts in many of the locations. Martin and his colleagues simply avoid any discussion of the fate of some 500,000 mammoths and the hundreds of thousands of crushed and splintered skeletons in Siberia and the Alaskan muck when ruminating on the fate of Pleistocene megafauna.

Another way to explain the extinction of the megafauna does exist, and we have seen that several Indian accounts suggest knowledge of and involvement with mammoths and other Pleistocene megafauna. If we were to simply cite these Indian traditions as evidence that some individual megafauna survived after the Ice Age and lingered on into nearly contemporary times, we would verify scientific interpretations but not add anything new to human knowledge. It would be better if new data could be obtained from the Indian memories.

Scientific knowledge progresses very slowly because people in science are reluctant to change any of their ideas until the evidence for new interpretations of data is so compelling as to make them seem foolish maintaining the outmoded doctrines. The Indian accounts of megafauna, while nebulous in terms of any measurable chronological scheme for dating, do contain considerable information on the environmental conditions that existed when these mammals were seen in North America by human beings. We will now turn to that additional information and try to determine what it tells us.

In chapter 5 I cited James Wright's report to John Bartram about Bartram's request that Wright secure bones from the Big Bone Lick in Kentucky. Radiocarbon dates for some of the bones and vegetal material recovered here are measured at 10,600 plus or minus 250 years. It does not seem possible that the bones could be that old. Could they have been lying on or near the surface for that period of time and not have disintegrated long ago? We should conclude, based on common sense, that the bones were reasonably fresh in the 1760s.

So let us examine the salient points that can be determined from the Delaware tradition. In order, the information that we can obtain from the Wright report is as follows:

1. Five reasonably complete skeletons, placed in a circle with heads pointing toward a common center.
2. Enormous bones with tusks ten or twelve feet long.
3. No such creatures had ever been seen alive by the Indians.
4. Legend said these creatures were once hunted by men of gigantic stature.
5. These giant men died before the demise of the animals.
6. "God" had destroyed the megafauna to protect the present race of Indians.
7. By implication, the Indians feared these creatures.

The information about past conditions and events contemporaneous with the mammoths that is interesting here is that giant men once hunted the megafauna; the gient men died out; and something eliminated the mammoths about the same time. Who were these giants?

Here we must pause and examine our vocabulary for a moment. Translators attempting to learn things from Indians, particularly at this time in American history, did not always render the proper meaning of what the Indians were saying. If presented with the assertion that coincident with these large creatures were tribes or groups of very large men, the chances are substantial that the translator would use the English word "giant" as an adequate substitute for the Indian word or description. The word "giant," unfortunately, has certain connotations in the English language and immediately suggests children's stories, folklore, and, in the minds of scientists, superstitions. But should the connotation be transferred with the translation of the Indian description?

From talking with elders of several tribes, my understanding is that the Indians were and are describing people of more than average height. In fact, some elders as a routine matter have reported that the Indians themselves were much larger and taller. A more general description of these people used by traditional elders is "the tall ones." This phrase implies cultural behavior comparable to the Indians in many respects, physical size being one important distinction that made these people memorable. Consequently, when "giant" is used in the discussions below, we should try to put ourselves within the conceptual universe of the Indians and not bring forward all the negative and fictional nuances that the English word carries. Since the vast majority of references to large people in published Indian legends use the word "giant," we are more or less forced to use it also. But we must understand that we are really talking about "the tall ones."

Giants and mammoths seem to have been part of another kind of world in which Indians were but minor, and sometimes helpless, players. We have seen this theme in many of the Indian traditions already discussed above. In the Pomme de Terre prairies the Indians hid from the large animals when they grew angry at the invasion of their hunting grounds by other creatures. The Beaver Indians told of a giant man killing a giant beaver, implying that some people, other than the Indians, were comparatively as oversized as the megafauna. Do other Indian traditions refer to giant-sized men who were contemporaneous with megafauna and with Indians?

The answer is "yes" and "not always." Ella E. Clark's collections of Indian traditions, *Indian Legends of the Pacific Northwest* and *Indian Legends of the Northern Rockies*, contain numerous stories of giant people. Unlike many writers handling Indian stories, Clark respected the elders who shared their stories and tried, as much as possible, to retain the exact wording and flavor of what they said, although, as we suspect in the translation of eastern Indian stories, she naturally shortcut the translations and used "giants" also.

In a story discussing the origin of the Chief's Face, a rock formation on Mount Hood, south of the Columbia River, an elder commented: "In those days [early times] the Indians were also taller than they are now. They were as tall as the pine and fir trees that cover the hills, and their chief was such a giant that his warriors could walk under his outstretched arms."[1] The mountain exploded, and the people could not live near it for a long time. When they returned to the area "... the children, starved and weak for so long, never became as tall and strong as their parents and grandparents had been."[2] The story predicted that the people would remain weak until a great chief came who could conquer the volcano spirit.

This tradition seems to be straightforward and appears to describe a condition of malnutrition which might be expected to occur if people were deprived of food for several generations. One would expect that after a century of living again in a fertile land the deprivations suffered as a result of the volcano would have been overcome and the people would have returned to a normal size again. We have seen a similar situation with the Great Depression generation in America. It is important to note here, however, that human beings, as a result of some change of living conditions, beginning with the eruption of a volcano, suffered an irreversible loss of size.

Clark also recorded a story of the Coeur d'Alene people involving giants. These men had a strong odor and apparently painted their faces black. They were "taller than the highest tipis" and "when they saw a single tipi or lodge in a place, they would crawl up to it, rise, and look down the smoke hole. If several lodges were together, the giants were not so bold."[3] Initially this description seems highly exaggerated, but again, if we know the Indian background, the story is not unreasonable.

Many tribes of the Pacific Northwest lived in pit houses much of the year. These houses were partially underground and partially aboveground, thus vulnerable to Peeping Toms from the outside. So to have these giant people looking through smoke holes does not imply that they had an unusual fictional height, but only that they were sufficiently tall so that they could easily look down on these kinds of dwellings.

Other tribes suggested that the giants herded the mammoths and otherwise used them as people in India today use elephants. There is no way that we can check this memory for accuracy. It would help to explain, however, the sparsity of mammoth bones in very recent times. Like domestic animals of the past century, domesticated mammoths would not leave a trace when they expired. A century ago everyone in most of North America used horses as a means of transportation, yet we do not find horse skeletons spread all over the landscape.

Some stories have retained just a faint memory that giants once existed. The time was long, long ago and a good storyteller of today will generally link these creatures with stories of strange animals and the "little people" because all three subjects have a tinge of the exotic in our present-day world. Whether they coexisted or not is another question. So I will not include these stories because they are suspect. On the whole, the trend of the stories seems to be that giants and our-sized people were contemporaries but did not have much contact with each other.

These stories of large people imply that Indians have a sense of historical sequence, somewhat different from Western history, but nevertheless of some significance because some unusual and spectacular observations and experiences are remembered for thousands, perhaps tens of thousands, of years. Would the Indian accounts be "accepted" by scientists today as evidence of the existence of giants and the very recent existence of some kinds of megafauna in North America? *Absolutely not.*

Examples of this bias are not difficult to document, and we should look at a case so readers can see the manner in which the Indian traditions are received. What happens when non-Indian scholars support the Indian version of events and suggest that it is historical?

In 1959, David Pendergast and Clement Meighan did field research in southern Utah with Paiute peoples living on the edge of modern civilization and certainly not familiar with much of the outside world. They published an article in the *Journal of American Folklore* entitled "Folk Traditions as Historical Fact: A Paiute Example," in which they compared information obtained from some Paiute elders with the data that other scientists had collected regarding a Pueblo-like people referred to as the Mikwitch who once inhabited the region.

Pendergast and Meighan devised the following classifications of knowledge by which to judge the oral tradition: (1) migrations and movements, (2) economic patterns, (3) physical appearance, (4) material culture, and (5) Paiute-Mikwitch relations. They demonstrated that there was a nice correlation between stories which had been passed down over the generations and what scholars believed they knew about these vanished people. They argued, therefore, that the Paiutes had memories of events and peoples that had been preserved for approximately eight hundred years.[4] We must note that Pendergast and Meighan might have been able to demonstrate substantially more correlations had they been members of the community, known *exactly* whom to ask, and been able to speak the Paiute language fluently. But their article is impressive in arguing the point of prolonged memories.

A rejoinder was published later, authored by Lord Raglan of Great Britain, in which he chastised the two Americans as if they were children forgetting basic facts. Raglan cited Edward Sapir's statement that "Indians believe myths and legend to be true" and then tartly commented that Pendergast and Meighan should have read the volume in which their article was published. There was, he claimed, "... ample evidence that Indian traditions do not distinguish between myth and history, nor between human and supernatural, ..."[5] on the basis that a statement from the Paiutes that some of these people became stones was sufficient evidence of the mixture of fact and fantasy.

It is important to note that Raglan did *not* examine the evidence presented by Pendergast and Meighan but simply cited a scientific doctrine, in effect using a "proof-text" to reject their presentation. "Turning into stones" might have been a Paiute way of saying "dead and buried." In fact, a good scholar, finding that everything else seemed to show a correlation, should have been alerted that "turning into stones" meant something different from the traditional English fairy-tale ending. It is almost impossible to get non-Indian scholars pried loose from their own cultural presuppositions to do careful interpretive work on Indian traditions. While they may loudly declare that the two cultural traditions are dissimilar, most of them do not seem to know what that really means.

Scholarly journals are littered with similar kinds of situations in which Indian accounts are rejected simply because they are Indian and theories are discredited because scholars are not willing to give credence to Indian accounts that seem to be accurate memories of times past. In general practice, a scholar cites an Indian tradition and then debunks it, classifying it as a psychological quirk illustrating the primitive mind. Lord Raglan's comments could have been equally applicable to Greek stories such as the *Iliad* and would have discouraged Schliemann from setting sail for Asia Minor.

Considering all the fraud that has occurred in scientific circles— from Kepler fudging his data, Mendel rigging his figures, Burt writing his own reviews of his publications, and, more recently, fraudulent reports in such areas as the breast cancer studies—it

seems ludicrous that a scientist would call into question the veracity of others under any circumstances. Add to that sorry history the factual worldview of the first Europeans who sought the
Seven Cities of Gold, Prester John's palace, and the Fountain of
Youth, a worldview from which the Western scientific tradition has
sprung, and it will be obvious that a long tradition of fantasy, myth,
fictions, and lies exists among Europeans and scientists that cannot be overlooked. Dismissing Indian accounts on the basis of predetermined doctrines, then, is not really a good idea and may
prevent us from retrieving some facts about Pleistocene America
that would be useful in understanding what happened over here.

Let us see what the overkill scientists say about the conditions
that greet and affect their alleged big-game hunters and then compare Indian traditions about giants and the demise of megafauna,
which would cover the same basic time frame. According to the
overkill thesis, the Paleo-Indians arrived approximately 12,000
years before our time. "Could a band of, say, 100 hunters," asked
Jared Diamond, "arriving at Edmonton breed fast enough to popu-
late an entire hemisphere in just 1,000 years?" Of course Diamond
believes they could and refers to the population rate of a modern
instance of settlers occupying new lands—the HMS *Bounty* muti-
neers who increased by 3.4 percent a year—as a situation that can
provide us with guidance. So Diamond argues, "At that rate, which
corresponds to each couple having four surviving children and a
mean generation time of 20 years, 100 hunters would multiply to
10 million in 340 years."[6] It would be even better to assume they
were all good old-fashioned Catholics and thus had twenty children for each couple.

With Indians dramatically expanding their population according to the overkill thesis, what is the result on the megafauna?
Again, referring to no megafauna except the mammoth, Diamond
suggests that "one quarter of the Clovis population was adult male
hunters who each killed a mammoth every two months." Diamond
then becomes poetic and mathematically determines that "six
mammoths [were] killed per four square miles per year."[7] There is

not the slightest shred of evidence of this kind of killing rate unless the Paleo-Indians were unusually successful in killing the mammoths without leaving any killing or butchering marks, or indeed leaving any skeletal remains at all. Fred Warshofsky, another advocate of the overkill hypothesis, elaborates on Diamond's idea thus:

> As the population of stone age hunters exploded, the game grew scarcer. The men pushed forward along a front that rolled from Canada to the Gulf of Mexico in 350 years and on to Tierra del Fuego at the tip of South America within 1,000 years. And then the disaster.
>
> After destroying all the big game on the two continents, man, the destroyer, was himself threatened with starvation. He was reduced to scratching the earth with sticks and overturning rocks to feed on grubs. As a result the human population swiftly declined as the major prey animals were hunted to extinction.[8]

So the scenario requires a population explosion from 100 hunters to 10 million hunters in 340 years, followed by a determined effort to track down the last individuals of the respective megafauna species on both continents. Since sex is usually a lot of fun, why didn't the Indians maintain the same birth rate into historical times? Why wasn't the continent as filled as China or India when the Europeans arrived? And why didn't these hunters then exterminate bison, elk, deer, and other animals?

So we have the spectacle of Paleo-Indian Clovis people who suddenly exterminated all megafauna and cannot or do not hunt the copious herds of bison, elk, moose, deer, antelope, and other herbivores which abound on the two continents. Now these formerly proud hunters eschew herds of herbivores and live on grubs from the underside of rocks. The Paleo-Indian population crashes significantly—and very conveniently for advocates of this theory. The population crash only leaves them to explain why so few Clovis sites remain to be examined after some 10 million Indians had been romping around the Western Hemisphere. *After all, 10 million people should have left a little graffiti and picnic litter somewhere.*

Some of Martin's colleagues try to support this scenario but give unsatisfactory answers. William Ellis Edwards wrote that "the most crucial factor in the extinctions here considered was apparently neither change in the physical nor in the biotic environment but in the sociocultural environment."[9] His reasoning is curious, to say the least: "... for many large game species, the only uncertainty was whether cultural evolution would enable a given species to be domesticated before being exterminated."[10] In other words, the Paleo-Indians are guilty of speciescide because they failed to domesticate the megafauna. Pray tell where in the world in recorded history have species been domesticated? Do we find examples of Africans domesticating the elephant or zebra, Europeans domesticating the Irish elk or mastodon? The argument is absurd.

At critical points like domestication the Indian traditions really do have something to contribute. There is a tradition that mammoths were herded like cattle. But the herders were giants! Reporting an old tradition, J.F.H. Claiborne noted: "There had likewise been a race of cannibals, who feasted on the bodies of their enemies. They ... were giants, and utilized the mammoths as their burden bearers. They kept them closely herded, and as they devoured everything and broke down the forests, this was the origin of the prairies. ..." And, he continued, "this cannibal race and the mammoth perished about the same time, by a great epidemic."[11] We thus have a scenario roughly comparable to the Big Bone Lick story and supportive of it from a tribe living several hundred miles away, in Mississippi rather than Kentucky.

Edwards, in an essay in *Pleistocene Extinctions,* made a valiant attempt to reconcile the disappearance of megafauna with explanations other than the big-game hunters even while basically agreeing with Martin. But his explanations are torturous at best. The most frequent explanation of the disappearance of these animals by nonoverkill scientists is that they failed to adapt to warmer climates. It hardly seems possible that any animal, living in a more benign region for a change, would promptly expire. As to the ending of the Ice Age, if the climate was so deleterious, the megafauna could simply have followed the glacial sheet north as it melted and today be prancing around the Arctic Ocean shores.

With scientists avoiding the climate-change argument as the element in extinction, that then required them to find another factor caused by extensive hunting that would explain the reduction of body size of the megafauna. We will follow Edwards's argument to the end to see if it holds up. "Human technology," Edwards writes, "including the use of missile weapons, greatly reduces the counterattacking defensive advantages of larger size and emphasizes concealment and speed of flight. At this point of increased pressure of human predation, the genetically selected optimum body size of many forms declines sharply."[12] While this explanation sounds scientific, it would also mean that all present game animals are declining in size because of human hunting activities, since there is certainly more pressure today than ever before in human history.

Accepting for the moment that Edwards has something here, what other effects would be possible? He finds that pressure by hunters "results in a decline in the average age and therefore in a nongenetic decline in the average size of prey," and therefore "with the downward shift in the age distribution, adults that mature faster and reproduce faster are genetically selected, and these tend to be smaller."[13] Apparently, if we follow this line of reasoning, megafauna were "selected" by hunters, not nature, because humans tracked the largest who were awkward and easy prey. The genetic makeup of the remaining animals was changed because younger individuals were now mating and began to produce smaller and smaller offspring. But the bulls or dominant males in the herds, flocks, or what have you, always had their choice of the younger females, although undeniably younger and virile males mated also. This argument just does not make sense nor is it based on any evidence.

One additional fact should be noted before we examine the Indian version of what might have happened. According to Edwards, "... human body size underwent a simultaneous postglacial reduction," and he believes this downsizing occurred "... primarily because of the shift to plant foods and the development of food preservation and storage associated with a more sedentary way of life," but adds, "... in part also because of the declining size of prey species and individuals during the Upper Paleolithic and Mesolithic."[14] The implication here is that these people lived on an almost exclusively meat diet, shades of the old view of cavemen, and that adding vegetables to their fare reduced their size. But Edwards's argument also suggests that as the game became smaller, the people eating it became smaller also. Does this mean that people who eat sardines are smaller than people who eat salmon? Again, we have absurd science offering bizarre explanations about things they do not know. Is it their status alone that makes our society take them seriously?

What we can establish, as common ground between science and the Indian traditions, is that many creatures, including human beings themselves, were much larger during the late Pleistocene and that body size decreased measurably. The real question, then, is whether these oversized creatures were destroyed or simply downsized to become present species and races.

We know that these large people were not destroyed by the Indians. Many traditions suggest a state of peaceful coexistence with the tall ones. Frances Densmore reports a Sioux tradition in her classic study *Teton Sioux Music*, which may give us some hints concerning this relationship:

> It is said that the thunderbirds once came to earth in the form of giants. These giants did wonderful things, such as digging the ditches where the rivers run. At last they died of old age, and their spirits went again to the clouds and they resumed their form as thunderbirds.[15]

Whether the giants dug the rivers or not, it would appear that they were on Earth when the major rivers in North America were carved out, and common sense would suggest that many of these rivers were caused by rapid glacial melting. But the Sioux seem to have admired the giants and considered them as positive people.

One tradition that does indicate conflict is that of the Indians near the old Spirit Lake at the foot of Mount Saint Helens. As the tradition goes:

> In the snow on the mountaintops above the lake, other Indians used to say, a race of man-stealing giants lived. At night the giants would come to the lodges when people were asleep, put the people under their skins, and take them to the mountaintop without waking them. When the people awoke in the morning they would be entirely lost, not knowing in what direction their home was.
>
> Frequently the giants came in the night and stole all the salmon. If people were awake they knew the giants were near when they smelled their strong, unpleasant odor. Sometimes people would hear three whistles and soon stones would begin to hit their lodges.[16]

The description of the giants throwing stones provides a great temptation to identify these creatures as Sasquatch, since this kind of behavior is said to mark some of their human encounters. But the giants could well have been the white-skinned race which forced the Salish, Sioux, and Algonkians out of the north country and then, if we follow Muller's thesis, migrated east and invaded western Europe, routed the Neanderthals, and are known as the Cro-Magnon peoples.

The demise of the megafauna, in almost every instance, is attributed by the Indians to an intervening act of the Great Spirit. With the tribes of Mississippi a specific cause, an epidemic which kills the giants is also mentioned. The act of the Great Spirit, given by many tribes as the cause of megafauna extinction, implies some kind of natural event that was understood as an act of grace by high spiritual powers.

The Sioux tradition does not simply talk about extinction but provides a context for the downsizing of the physical shape of the animals and so is worth examining. As told by Charles Eastman, the story is that the land was well populated with Indians and they got along well with the animals. It is said, not only by the Sioux but by many other tribes, that the animals and humans could talk to each other. But, as time passed, the buffalo and other grazing animals got out of hand and "made war" on the other peoples, including the Indians.

So the Great Mystery made a great tent and kept it dark for ten days. "Into this tent he invited the different bands, and when they came out they were greatly changed, and some could not talk at all." The animals were stripped of some of their physical attributes and talents. "The buffalo came out of the darkened tent the clumsiest of all the animals. The elk and deer were burdened with their heavy and many-branched horns, while the antelope and deer were made the most defenseless of animals, only that they are fleet of foot. The bear and the wolf were made to prey upon all the others."[17]

Many of the Sioux believed that the event had centered in the Devils Lake area of North Dakota and held the place to be sacred. It seems likely that the "dark tent" was some kind of traumatic climatic activity wherein the sky turned black for a significant period of time and great changes took place on the Earth. Whether the animals were changed instantaneously or not, the story suggests that animals were once the dominant life form in the Plains area and some major event reduced them in size and human beings were then better able to deal with them.

William Tall Bull, Cheyenne elder, said that there once was a giant buffalo that was carnivorous and greatly oppressed the people, but the Great Spirit reduced him in size. Thereafter the Cheyenne would not eat a certain kind of fat found in the throat of *Bison bison* because it represented human flesh that the giant buffalo once ate. Both the Sioux and Cheyenne insisted that the two species of buffalo were actually the same animal, reduced by this strange intervention of the Great Mystery.

Loren Eiseley commented on the idea that the two bison were related and rejected the idea, although not knowing at the time that the Indians believed it. When refuting the original thesis of overkill by fire drive, Eiseley wrote: "... I cannot vision *Bison bison* as a trivial mutation appearing casually at the scene some 2,000 years ago and multiplying quickly after the demise of an earlier fauna exterminated, so some would have it, by dextrous use of the spear thrower."[18] But the La Brea skeletons seem to support the Indian account. Chester Stock writes that "the systematics of North American bison species is currently not well understood, although the migration of the genus from Asia to North America characterizes the beginning of the Rancholabrean Land Mammal Age. It seems entirely possible that *Bison latifrons* constituted the ancestral form from which the smaller *B. antiquus* evolved. Indeed, Lundelius et al. [citation omitted] suggested that *Bison antiquus* may represent a smaller-horned end member of the *B. latifrons-B. alleni* lineage."[19]

Indeed, the Indian version does not have *Bison bison* as a trivial mutation conceived along the endless periods of time which evolutionists need to create a species. Rather, *Bison bison* is the direct descendant of the larger bison species as altered by some spectacular physical event. If we were to follow orthodox scholarship, it would be necessary to have giant-sized bison migrating across the Bering Strait around 15,000–20,000 B.C. and have them mutate into a smaller version in time to be entombed in La Brea along with the larger species around 11,000 B.C. This time span for bison evolution would be ridiculously short.

What if *Bison bison* was not the result of genetic mutation but was the identical species of animal living in a substantially altered atmosphere that did not encourage growth to mega-size? Donald Patten, in a highly informative and perceptive article entitled "A Comprehensive Theory on Aging, Gigantism and Longevity," raised a series of important questions about megafaunal size and the subsequent reduction to contemporary species size that bear reviewing. I will paraphrase most of his argument, since it is precise, detailed, and at times technical. The importance of his suggestions cannot be minimized because they deal with factors which are conveniently avoided in the discussion of megafauna extinctions.

First, Patten points out that gigantism occurred across a great many species and consequently the large animals, birds, and reptiles we see in the Quaternary were not confined to the Western Hemisphere but were found on all continents during a certain period of Pleistocene time. Even if we reject the evidence and arguments offered previously and insist on placing blame on the Paleo-Indian hunters for loss of megafauna in the Western Hemisphere, we still need to explain the complete loss of megafauna and megaflora on the other continents. Why did these fauna grow to be king-sized in the first place, and why did some species not remain until historical times? Or did they? Did some individuals of mega-species survive on all continents until human contact, giving rise to the legends of monsters? Are ancient tales about horrible monsters a reflection of surviving megafauna? Were some creatures still extant in England, allowing the Knights of the Round Table to spend their time slaying them in the Maiden Protection Program initiated by King Arthur?

Second, Patten suggested that most of the megafauna could have vanished over several generations because of a radical change in environmental conditions and that if we lined up the various sizes of a particular species, we could see that they are really the same creature, only in a decreasing series of sizes. Patten uses as examples the different species of bison in North America and the various species of large birds in Australia and New Zealand. If, he argues, we arrange these species in a descending order of size, we would be describing the effects of a drastic change of atmosphere on a single species of animal. Considering the recent discovery of dwarf mammoths in Alaska, this theory has much to recommend it.

Patten links the propensity toward gigantism with longevity, suggesting that the two phenomena seem to occur together, and that at one time animals and people grew larger than today and lived longer. He points to the advanced ages of the patriarchs recorded in the Bible and references to "giants in the earth" and "mighty men" as evidence that the Near East had its gigantism also.

Patten demonstrates in several writings that in the preflood era the people regularly lived unusually long lives. It thus was not uncommon for men in their seventies and eighties to be fathering children and for many generations to have shared long periods of time with each other.

Patten creates a graph that startlingly demonstrates his thesis. As we approach modern times, human life spans begin to approach the shortness we know today. Once human longevity stabilized around four score for a life span, people began to doubt that anyone ever lived much longer. Even in ancient times extreme skepticism existed regarding longevity. Patten cites the efforts of Josephus to convince his readers that people really did live longer in ancient times. Josephus, in mentioning the advanced age of people before the flood, argued:

> Now I have for witnesses to what I have said all those that have written Antiquities, both among the Greeks and barbarians, for even Manetho, who wrote the Egyptian history, and Berosus, who collected the Chaldean monuments, and Mochus, and Hestiaeus and beside these, Hieronymus the Egyptian, and those who composed the Phoenician history, agree with what I here say: Hesiod also, and Hecataeus, Hekkanicaus, and Acusilaus, and besides Ephorus and Nicolaus relate that the ancients lived a thousand years. ...[20]

This list of authorities, already ancient in Josephus's time, is certainly impressive, and instead of shucking it aside as evidence of the superstitions of the ancients, we might learn something by taking it seriously. Many times Indian elders have told me that in the old days people lived until they were two hundred years old and, while it sounds unlikely, there is no good reason to doubt them. I have encouraged some of them to talk openly of these things but they are reluctant to draw the fire of skeptics.

So we are talking about a golden age when there was very little hardship, when there were considerably more species of animals living on all continents, and when there was no Ice Age. This idyllic planetary condition is remembered in a substantial number of human societies but, according to Patten's theory, was shattered by the appearance of a comet or meteor composed almost wholly of ice and water which passed close enough to the Earth to disintegrate, dumping ice in massive amounts on the magnetic poles and precipitating an ungodly amount of rain on the temperate regions.

Large amounts of ice quickly formed gigantic icy mountains in areas of the Northern and Southern Hemispheres where we had glaciers, producing, as some scientists have speculated and as the Shoshones reported, mountains of ice that reached to the sky. The impetus of this dump provided the mechanism to enable the ice to travel in all directions, most particularly uphill, at a very rapid rate of speed, accounting then for the glacial "advances" and, if there were several dumps over a short period of time, the "retreats" or "stages" of glaciation as well.

Prior to the disaster, Patten suggests, the carbon dioxide in the atmosphere was considerably higher than our present level. With the icy catastrophe, the oceans cooled significantly, and as the temperature of the water decreased, much of the CO_2 was absorbed by the colder water. Some carbonate strata were formed—limestone, dolomites, and dolostones—and more CO_2 was taken out of circulation. About 50 to 75 percent of the biomass was buried, fossilized, and deposited in strata as tidal waves laid down immense beds of plant and animal matter. Patten writes that "... we can conservatively defend 15 times the current concentration which suggests 5,000 parts per million, or ½ percent. If we were to speculate liberally, a 2 percent mixture of carbon dioxide in the pre–Ice Age atmosphere could be proffered, but it is probably high."[21]

Several studies have shown that carbon dioxide enhances plant growth and Patten suggests that all organisms are somewhat affected by the amount of carbon dioxide available to them. The hypothalamus gland serves as a supervisor of the hormone system and carbon dioxide acts as a stimulator for it. An excess of carbon dioxide, at least something greater than our present concentration in the atmosphere, would stimulate cerebral circulation and oxygenation. Too much would tend to act as an anesthetic. In preflood days, then, there was sufficient carbon dioxide present to create gigantism and promote longevity. Reduction of the percentage of carbon dioxide meant slower growth and a downsizing of the fauna and flora.

We actually hear about CO_2 every day but in a different context—global warming and air pollution. During 1995 there were constant alarms by scientists concerning the breakup of the Antarctic ice shelf with icebergs the size of Rhode Island breaking from the main body of ice. Our atmosphere is definitely getting warmer and having a drastic effect on the polar ice sheets. In a sense we are returning to prediluvian conditions.

We never stop to think that there were once corals growing in northern Norway and lush foliage in Alaska and Siberia. Our fossils give evidence of a much warmer Earth. And this warmer Earth had megafauna, suggesting that opportunities for maximum growth were present also. Among other factors, this warm climate has supported substantially more plant growth, which must have produced increased CO_2.

The Earth's atmosphere, then, might have originally been much different and possessed maximum benign living conditions. With a maximum benign percentage of carbon dioxide during much of the planet's history, producing monstrous-sized dinosaurs would not have been difficult, given carbon dioxide's effect on growth. The coefficient of gravity would have been somewhat different and perhaps the Earth, minus the tremendous amount of ice and water dumped on it, might have been a different place to walk, run, and live. A collapse of that atmosphere once the dinosaur age ended and again in the Pleistocene would significantly reduce the size of animals over several generations until they were again adjusted to the present atmosphere. Every living creature would be affected and that would be why, according to the Indians, these giants are simply "gone."

Derek Ager, the dean of European stratigraphers in geology, summed up the theories purporting to explain the major extinctions of the geologic past: "Almost all the theories (including the Noachian one) that seek to explain major extinctions in the past, lead by one route or another to climatic oscillations and related matters such as the composition of the earth's atmosphere. These in turn tend to point to extraterrestrial phenomena."[22] Patten certainly suggests an extraterrestrial source of planetary change, but he does not rest there. He takes seriously the effect this source would have on the atmosphere and does not simply try to locate mechanical geological disruptions.

The problem with orthodox interpretation of the relationship of the megafauna to creatures of our present size is that most scientists have looked for *genetic* change, as the quotation by Loren Eiseley above demonstrates. They have therefore constructed a large bestiary of megafauna, and dinosaurs perhaps, which have no ancestors and no descendants. Instead of grouping animals by similarity of form and considering that they may represent a single species varying its size in accordance with the manner in which they were organically stimulated to grow, they have created all kinds of species.

Again, orthodox comments regarding the animals at La Brea are relevant here. According to Chester Stock, "... the puma and lynx occurring at La Brea are closely allied to types still living in western North America. The living species of puma (*Felis concolor Linneaus*) is included in the collection, and the extinct *Felix bituminosa Merriam,* and is now generally considered to be indistinguishable from the living puma. The lynx or bobcat, from Rancho La Brea, originally assigned to an extinct subspecies (*Lynx rufa fisheri Merriam*) on the basis of its cheek teeth, is no longer considered distinct from the living lynx."[23] What seems to have been happening at La Brea is that scientists are abandoning fictional species classifications and admitting that a larger size does not necessarily constitute another species.

If we look carefully at other scholars we will see a trend in favor of abandoning the multiple classifications and admitting that megafauna in many cases are not genetically different from their smaller, later relatives. Thus E. A. Vangengeim, in an essay entitled "Quaternary Mammalian Faunas of Siberia and North America," writes that "a remarkable similarity is observed in the development of these forms [mammoth and bison] in Eastern Siberia and Alaska during the second half of the Pleistocene: a gradual dwarfing resulted in dwarf mammoths and smaller-sized bison with shorter horns."[24]

The response of many evolutionists is that the geological strata show a progression from small to large. The strata do indeed show this progression because they have been arranged to demonstrate precisely that very point. The strata are generally identified by "index fossils," and the choice of index fossils is somewhat arbitrary. We do not have a record of how many times scientists have changed index fossils when their theories became untenable. Quite often there are as many small fossils as large ones of any species in any strata of a geological column, belying the idea that there was organic progression. When talking about climatic change, scientists should have been looking at the types of support that environmental and atmospheric changes gave that either inhibited or promoted growth of organisms. In other words, as organisms breathe and/or use air, the composition of the air makes a radical difference in the maturation process, the size of the organism, and its longevity.

Could such a change have happened? It certainly sounds reasonable to me. When I was a child the size of athletes was considerably less than it is today. Indeed, if your high school basketball team had anyone over six feet tall, the opposing team would accuse your coach of bringing in a "ringer" and demand to know whether the boy's family had been given a job or promised special favors in order to get him to enroll at your high school—at least people in rural South Dakota were sensitive to these things. Today, boys in high school are frequently over seven feet and in junior high we are now seeing some people six-and-a-half-feet tall.

Some college football teams have offensive and defensive lines that on average approach 300 pounds per player. These significantly larger athletes cannot all be using steroids, nor can the increase of salaries for power forwards and tight ends have created the size of the players we have today.

Simply compare the size of the medieval knights' armor, Civil War army uniforms, and modern athletic equipment to see what has been happening. In 1735, Jean-Baptiste Le Moyne, sieur de Bienville, wrote to the French Colonial Department in Paris complaining about the quality of soldiers being sent to help him govern Louisiana: "There are but one or two men among them whose size is above five feet, and as to the rest, they are under four feet ten inches."[25] Our modern athletes would be regarded as giants by these French soldiers of less than three centuries ago. During the last half of this century we have significantly increased the amount of carbon dioxide in our atmosphere and we now have "Refrigerators" playing football. How would our basketball players be described in the vernacular if we were able to transport them back to King Arthur's court? Giants?

Actually, the size of human beings started to increase with the advent of the Industrial Revolution and the introduction of more carbon dioxide into the air. Patten cites some startling figures: "With the onset of the petroleum age, atmospheric CO_2 began to increase even more sharply. In 1957 the concentration of Hawaii was 311 ppM, adjusted to seasonal variations. The same reading was taken at the South Pole. By 1971 this had risen to 322 ppM. The rate of increase during the early 1970s in fact has been steepening slightly. By extension the CO_2 concentration by 1979 is 329 to 330 ppM. The concentration of atmospheric CO_2 is increasing between .7 and .75 ppM per year."[26]

A related factor here is how an increased percentage of carbon dioxide in the atmosphere would affect our radiocarbon dating, which has all of the megafauna clustering at around 12,000 years ago. If there was significantly more carbon in the atmosphere, the initial premise of radiocarbon dating—determining the amount of carbon-14 in vegetal and organic material—would be much different at its starting point. We could not assume, as we do today, that the percentage of carbon in material in the late Pleistocene was the same as what we find today. It might be considerably more or it could be considerably less. We simply have not thought about these things.

Scientists casually say that climatic conditions have changed, implying a rise or fall in the surface temperature of the planet, but then they do not think through the remainder of the problem. Since carbon dioxide is absorbed in cold water rather easily, the onset of the Ice Age, even under the inconsistent orthodox theories, would have reduced the amount of carbon dioxide in the atmosphere simply by decreasing the mean average temperature of the oceans. Thus, whether Patten's more comprehensive theory is accepted or not, adjustments have to be made in scientific thinking regarding atmospheric carbon dioxide.

Change of atmospheric composition, then, is a reasonable solution to the question of gigantism and the demise of the megafauna of the Pleistocene. Watching this process of downsizing, Indians may have felt they were being saved by the Great Spirit. They may have seen individuals of both sizes in the same area and attributed the change to radical activities within the "dark tent," or they may have attributed the demise of these creatures to an epidemic. Indians themselves may have been reduced in size, not through malnutrition but because of the change of atmosphere. The explosion of Mount Hood would have been a minor local event, a mere hiccup within the Ice Age disaster. But effects of the new atmosphere would have caused a permanent reduction in size that the Indians recognized.

Considerably more research and thought must be done before any alternative theory can become a rival to the entrenched orthodoxy concerning the events of the Pleistocene. The problem is in getting the advocates of orthodoxy to admit that there is a problem, that the solutions they have suggested are not really convincing, and that they cannot continue to offer ad hoc answers to pressing questions. No advances can be made in our understanding of our planet and ourselves when scholars simply recite doctrinal beliefs as if they could verify sagging credibility. Science increasingly acts as if it were a religion by relying on authority rather than argument of evidence. In the next chapter we will look at another field of knowledge in which the Indian traditions have a divergent viewpoint and interpretation.

8

Geomythology and
the Indian Traditions

IN CORRECTING INADEQUATE THEORIES such as the Bering Strait migrations or the big-game hunter "overkill" we inevitably come into conflict with the prevailing scientific paradigm. The points of conflict spread out from anthropology to archaeology and then to paleontology in a disconcerting manner. Evolutionary biology, as represented by the field of paleontology, has been in an incestuous relationship with geology ever since the Darwinians wrested the symbols of authority from the theologians. Therefore, as we raise questions about glaciation, land bridges, and the disappearance or transformation of megafauna, we are led into an examination of some aspects of geology.

The methodology of geology is to arrange geological strata by first identifying index fossils and then classifying the fossils from smallest and simplest to creatures with more complexity until we reach the most highly developed organisms, in accord with the dictates of evolutionary theory. Once that sequence is arranged, evolutionists then enter the picture, point to the geological stratigraphic column illustrating the progression from simple to complex, and proclaim that evidence exists that evolution really occurred.

It has been very difficult for anyone to get "inside" this fortress of the two incestuous disciplines—evolutionary biology and geology—and raise the relevant questions about either evolution or the stratigraphic column. Yet it is suspected that the stratigraphic column and evolutionary family trees are largely the figment of scholarly imaginations, but highly orthodox imaginations.

Many puzzles abound in these sciences. The Grand Canyon, for example, on the Colorado Plateau, is located in one of the most stable areas of the planet geologically, and yet as we go down its strata we discover that it lacks the Upper Carboniferous (Pennsylvanian) period and the Ordovician and the Silurian systems.[1] And if we catch evolutionists at an odd moment, they are hard pressed to provide us with even one good example of an evolutionary chain of any species. Thus, the seemingly endless periods of time preceding the present are not as well documented as we might suspect. They are necessary only because both geologists and paleontologists need immeasurable amounts of time to invoke imagined incremental evolutionary changes.

Fortunately for us, one of the insiders in the geological establishment, in the closing years of his life, began to write highly unusual books asking questions which geologists and evolutionists had refused to ask themselves for over a century. Derek Ager, in two startling books, *The Nature of the Stratigraphic Record* (1973) and *The New Catastrophism* (1993), asked a series of questions that in his opinion had never been satisfactorily answered by his discipline.

The first book gave great comfort to catastrophists and fundamentalists. Because of the spectacular nature of the complaints from orthodox scholars, Ager emphatically announced in his second book that he did not disbelieve evolution or regard his work as providing a framework for verifying the existence of Noah's flood. But there can be no mistaking the scope of Ager's discontent and bewilderment with orthodox geological interpretations of strata or his willingness to go beyond mere commentary and provide for his readers and for future scholars an arsenal of facts which could be used to critique the traditional doctrines of these fields.

Three basic points emerge in Ager's survey of geological doctrines that have relevance for our critique of how Indian traditions have been handled by scientists. First, there is a real discontinuity between doctrines describing the formation of strata of any kind that we find anywhere in the world and the actual situation. Second, many volcanic formations, although dated very early in geological time in the orthodox scheme, have a disturbingly modern appearance, suggesting that something is amiss. Third, Ager sometimes suggests that early and/or primitive man may have been an eyewitness to geological events of some magnitude in different parts of the globe.

My purpose, in moving the discussion now to the field of geology, is to offer evidence from the Indian traditions that would suggest that American Indians have occupied the Western Hemisphere for very long periods of time and could not have been latecomers to this continent via the Bering Strait around 12,000 years ago. In order to present this evidence I will first look at some of Ager's critiques of the present state of geological knowledge and then examine traditions recounting what the Indians say occurred a long time ago.

Geology, like all other scientific disciplines, suffers from ethnicity. That is to say, while we like to think of science as one big happy family, there is great competition among and between the scholars of the different countries. Since each national scientific establishment is an authority of its own—within its national borders—discoveries often tend to particularize themselves. Thus, geologists of England classify strata with names honoring Englishmen or English locations, and the same strata in France will have French nomenclature. German, Austrian, and Russian geologists will have their own names for local strata. Geologists, and by extension the interested public, begin to think that the stratigraphic column is several miles high when all strata are in theory lined up in a proper sequence. But, as it turns out, if the diversity of national names is eliminated, the stratigraphic column becomes much simpler and suggests some startling facts that were previously hidden by the ethnicity.

Ager registers his concern about the manner in which knowledge of our planet's past has been coordinated. He then reviews what we think we know about certain geological strata, beginning with Upper Cretaceous Chalk. This formation constitutes the popular White Cliffs of Dover but is also found from "Antrim in northern Ireland, via England and northern France, through the Low Countries, northern Germany and southern Scandinavia to Poland, Bulgaria and eventually to Georgia in the south of the Soviet Union." But, he notes, it is also found in Egypt and Israel, and in Texas, Arkansas, Mississippi, and Alabama in the United States, and finally in Australia.[2] Under present theories about how sediments are deposited, it is not possible to explain the widespread occurrence of this particular strata.

Ager moves on to discuss the Coal Measures and states:

> Whatever the vertical and lateral changes in the Coal Measures, we still have to account for a general facies development in late carboniferous times that extends in essentially the same form all the way from Texas to the Donetz coal basin, north of the Caspian Sea in the U.S.S.R. This amounts to some 170° of longitude, and closing up the Atlantic by a mere 40° does not really help all that much in explaining this remarkable phenomenon.[3]

And again with the great continental red sandstone of the Devonian period, which contains the extensive fish remains, the formation is spread across Canada, throughout Europe, and into Kashmir.[4] Ager continues on and examines other kinds of strata. Limestones, gravel conglomerates, and other formations are shown to extend over exceedingly wide geographical areas. In fact, if the different scholars did not zealously protect their own strata, we would have an exceedingly simplified stratigraphic column that might reduce substantially the estimate of geological time.

The problem posed by these strata is that they suggest a blanketing of the planet from extraterrestrial sources. Ager, of course, does not directly advocate this explanation; perhaps he wished to maintain himself in good standing with his profession. But one cannot begin to imagine the size of the mountain, or extent of the ocean, that would be necessary to lay down widespread deposits of this magnitude. Sedimentary rocks are something more than we have been taught.

Regarding sedimentation, Ager notes that "even in such classic areas as the Mississippi delta, where sediment is thought to be accumulating rapidly, there is plenty of evidence to suggest that, after building up for a while, much of it is carried away again."[5] And he writes that "if we look at the sea-floor maps that are now becoming increasingly available, one is struck (at least, I am) by the great areas that are either receiving no sediment at all or else are covered with the merest veneer."[6] In summary, we cannot begin to explain the origin of what we have called sedimentary strata because there are no processes of deposition that we can observe that would create anything resembling what we see in rock formations today.

Various volcanic formations also come under Ager's critique. Citing "... the Variscan massifs of Europe, such as the Spanish Meseta, the Massif Central of France, the Eifel of Germany, the Bohemian Massif of Czechoslovakia, and the Rhodope Massif of Bulgaria," Ager suggests that they have "... all the features of volcanicity that lasted late enough to terrify Paleolithic man and perhaps provide him with his fire. Some of the craters look so fresh that one almost expects the rocks still to be warm."[7] Ager also mentions the black basalts of Arizona, commenting, "These too date from the Tertiary, but have stayed remarkably fresh thanks to the arid climate of the American southwest.[sic]"[8] It is all the more puzzling to have fresh-looking volcanic formations in the Southwest protected by dry climate, as we are told that the spectacular landscape—including Ship Rock and Monument Valley—are the result of extensive erosion caused by winds and dry climate. Although he does not specifically describe basalts everywhere in the world, the sites he does mention all seem to have fresh volcanic rocks that have not weathered or eroded since they were laid down.

Finally, Ager is among the few scholars who have suggested that some geological events may have been experienced and remembered by early human beings. Discussing the very recent volcanic areas within the Carpathian bend in Romania, Ager speculates that "it may well be that early humans first kindled their fires from the conflagrations caused by such geologically recent eruptions."[9] And again he suggests: "... many early humans must have seen geological phenomena far more violent and spectacular than any we know in historic times, including the last great volcanicity across northern Europe from the Auvergne to Romania and the Explosion of Santorini which may have given rise to the Atlantis legend. In New Zealand the first Polynesian immigrants may have seen and suffered some of the last huge volcanic explosions in North Island."[10] Philip King echoes this possibility in his book *The Evolution of North America*, a rather technical treatise which gives a comprehensive overview of the geologic history of our continent, when he notes that "... all the evidence seems to point to the astonishing conclusion that the Grand Canyon was cut largely in the two million years or so of Pleistocene and later time, or a time when men were living already on the earth. If primitive man had been living in North America, he would have witnessed the formation of the Grand Canyon."[11]

There are indeed many traditions of human societies relating times when the Earth boiled and fumed, when, as the Hopis say, the world was destroyed by fire. If some geologists are now speculating on the possibility that people saw and remembered such events, why aren't these accounts useful in science? Here the arrogant attitude of scientists that all early people were frightened of nature and formulated fictional tales to explain the origin of things precludes scholars from using these accounts.

Scholars have devised a technical language to deal with the traditions of the past and non-Western peoples, and this language is designed to cleverly divert this nonscientific information into harmless categories where it cannot disrupt the doctrines we are currently supporting. We have already seen how Martin simply rejected carbon-14 dating that didn't agree with his thesis and how scholars attacked any notion of historicity within the traditions or memories of Indians regarding the past and their experiences with megafauna.

Three basic concepts stand in the way of examining the traditions of Indians in a fair and intelligent manner: "myth," and its progeny "euhemerism" and "etiology." *Myth* is the general name given to the traditions of non-Western peoples. It basically means a fiction created and sustained by undeveloped minds. Many scholars will fudge this point, claiming that their definition of myth gives it great respect as the carrier of some super-secret and sacred truth, but in fact the popular meaning is a superstition or fiction which we, as smart modern thinkers, would never in a million years believe.

Within the broad classification of myth are two subcategories of story-line creations: "euhemerisms" and "etiological" myths. The *euhemerism* is a narrative which contains some participation of the supernatural that is wholly constructed by primitives and which they insist is historically true. For decades the Trojan War was believed to be a euhemerism until Heinrich Schliemann began to dig tells in Asia Minor and proved the conflict to have a historical basis. An *etiological* myth is a narrative made up to explain something which people have observed or which they wish to explain in familiar terms. Looking at various kinds of landscapes, in the etiological format we simply assume that primitive and ancient people would make up a story, based on their knowledge of nature, to account for waterfalls, volcanoes, rivers, and so forth. Most of modern science is, in fact, etiological myth, since we cannot explain fossils, we cannot explain sedimentary deposition, and we cannot explain the causes of glaciation.

It is possible to separate non-Western traditions from the mainstream of science and keep them comfortably lodged in the fiction classification because most of them contain references to the activities of supernatural causes and personalities and are not phrased in the sterile language of cause and effect, which has been the favorite language of secular science. It is unfair to do so, however, when scientific writers have complete license to make up scenarios of their own which could not possibly have taken place and pass them off as science and therefore as superior to other traditions.

Consider, for a moment, the improbable scenario devised by Jacob Bronowski to explain the origin of bread wheats. This is the situation which precedes his explanation. All previous grasses have been able to separate the grain from the ear and all are, of course, fortunate mutations that by chance work their way toward human edibility. Suddenly a new bread wheat mutates and grows but it cannot separate its ears from the stalk. Fortunately, when the first plants of this new bread wheat begin growing, humans happen by. So, Bronowski says:

> Suddenly, man and plant have come together. Man has a wheat that he lives by, but the wheat also thinks that man was made for him because only so can it be harvested. For the bread wheats can only multiply with help; man must harvest the ears and scatter their seeds and the life of each, man and plant, depends on the other. It is a true fairy tale of genetics, as if the coming of civilization had been blessed in advance by the spirit of the abbot Gregor Mendel.[12]

This explanation passes for science while it is admittedly and unabashedly a fairy tale. Does the wheat really have thought processes which enable it to sigh in relief when men begin harvesting it? Scientists may cry "unfair" and argue that Bronowski should be allowed a little leeway in his description. Granted, but why should scientists be allowed leeway and others denied it? All we ask is respect for the other traditions and some of their versions of origins. Let the use of "spirits" in non-Western traditions be the equivalent of genetic fairy tales in scientific thinking.

In the early 1970s, Dorothy Vitaliano attempted to show that some information possessed by ancient peoples and the non-Western tribal groups, and classified as "myth," might indeed be useful. She began to match some accounts with modern geologic knowledge to create a new discipline which she called *geomythology*. Geomythology, according to Vitaliano, is an effort "to explain certain specific myths and legends in terms of actual geologic events that may have been witnessed by various groups of people."[13] In a very real sense, linkage of traditions and legends with present-day knowledge might provide some additional data for scientific experimentation; it would also verify the historical basis of the legend, take it out of the category of folklore, and give it some real status.

Minimally, verification would suggest that a particular group or tribe of people lived at a specific location and had been witness to a geological event. In an unexpected format, Vitaliano was stepping forward to provide substance to Ager's offhand remark about the geologic events that the ancients might have experienced. It should be apparent how useful this approach is to American Indian efforts to get traditional knowledge verified.

Since our knowledge of prehistory in every continent is a matter of best guess by a prominent scholar and slavish acceptance of his idea, at least until his retirement or demise, geomythology promised to give us additional tools to probe the past. But it is not without practical application in the present, and this ability to provide additional data for the solution of real problems will perhaps keep geomythology from emerging as a popular field of study for several more generations.

Litigation has already been conducted in which geomythological evidence would have been critically important. Several years ago a human skeleton was discovered within the lands claimed by the Shoshone and Bannock tribes in eastern Idaho. Graciously they allowed some scientists at the University of California to conduct some tests on the bones and, after a decent period of time, wanted the bones back for reburial. The scientists resisted the return and the case went to court. One of the most rabid scientists argued that no one except the queen of England could know where their ancestors' bones of even five hundred years ago were buried, hence the skeleton could not have been a Shoshone or Bannock and they had no claim to the remains. At this point, geomythology could have been critical because there is a Shoshone tradition which begins: "When the Big Horn basin was a sea," indicating the possibility of a rather prolonged tenure by Shoshones and Bannocks in the Wyoming/eastern Idaho region.

Vitaliano's treatment of Indian traditions is puzzling, at least in my mind, because she allows some possibilities which seem remote while denying others which seem convincing. Let us examine one of her examples and then examine a variety of sites and see if we can establish any body of evidence that would suggest eyewitness accounts and the transmission of these stories to present-day Indians, at least to the Indians who were first questioned about their memories of certain geologic locations and events.

THE BRIDGE OF THE GODS

Several Indian tribes in the Pacific Northwest have traditions regarding a geological formation which they say once existed west of The Dalles dam on the Columbia River. One version of the story suggests that the Columbia River once went underground, presumably as it passed below the Cascade Mountains, finally emerging near the coast. This phenomenon is not unusual, since the Humboldt River "sinks" in several places in Nevada as it moves west toward the Sierra Nevadas.

In 1921, a very old Wishram woman, well over a century old, who could remember when Dr. John McLoughlin established Fort Vancouver in 1825, told her tribe's story of this formation. The underground tunnel was frequently used by the Indians to avoid climbing the Cascades when traveling to the Pacific Ocean. "Whenever a party of Indians reached this long tunnel," she said, "they would fasten their canoes together, one behind the other, so that they would not crash against each other in the darkness. Then they would pray to the Great Spirit for courage and guidance as they paddled through the long, dark tunnel."[14] Some scholars, including Vitaliano herself, have expressed skepticism about this geological formation, claiming that the sides of the present

Columbia gorge do not indicate the possibility of any bridging structure having been there.

Many tribes have stories about the Bridge of the Gods and when these accounts are compared, the Wishram version seems to be the earliest. The most frequently repeated narrative suggested that Mount Hood and Mount Adams quarreled (usually over a maiden, since all old tales are eventually romanticized) and began to hurl hot rocks at each other. The conflict became so intense that the bridge over the river collapsed, in effect freeing the river from its underground course and creating the present-day Columbia.

Vitaliano suggests that an earthquake was involved and dumped a massive amount of material into the Columbia to form The Dalles. Geologic evidence suggested that there was once a giant landslide between Table Mountain and Red Bluffs which did block the Columbia. I suspect that a couple of places along the Columbia could qualify as the location and that all versions of the story refer to one or the other site.

Vitaliano had some difficulty interpreting the Indian time scale, since some of the versions suggested that the event had taken place in the time of their "grandfathers," which she dated to mean between 1750 and 1760. "But," she argued, "from the geologic evidence, the landslide could have happened as much as a thousand years ago. ..."[15] The problem is that when Lewis and Clark passed by this location, large trees standing upright with their branches could still be seen some thirty feet below the water. It would seem unlikely that trees could remain for nearly a thousand years without some disintegration. Vitaliano concluded, "Had the event occurred as recently as the middle of the eighteenth century, I feel certain the tradition would probably reflect the geologic facts somewhat more closely than does a mythical bridge. As it is, except for the implication that the Indians witnessed some activity of Mount Hood and Mount Adams, the Bridge of the Gods, like the explanation of The Dalles, seems to be a purely etiological invention."[16] But would the Indians have devised a complicated story of a dark tunnel under the Cascades and would so many tribes have preserved their own version of the bridge unless it had once been a prominent landmark in the region?

Surely the Indians had seen the Cascade volcanoes erupt. If we can only suggest that they marked the occasion of two volcanoes erupting simultaneously by making up a story about a bridge across the Columbia River, which would have no connection whatsoever with volcanoes no closer than fifty miles, what possible motivation can we suggest? Combining many geological formations in an etiological myth might be a way to deal with the *fact* of creation. But it seems unlikely that so many tribes would put together the same basic narrative about the site unless there was some reality behind the tale.

Therein lies the difficulty in approaching the oral traditions of Indians from a Western scientific perspective: instead of postponing judgment and viewing the anomaly as a prospect for future research, conclusions are drawn prematurely, are almost always in favor of rejecting the Indian account, and the usefulness of the tradition is lost. Instead, we are given doctrinal assurances that Indians made up the story.

MOUNT HOOD

One of the two volcanoes cited in the Bridge of the Gods story, Mount Hood, is mentioned favorably by Vitaliano as an instance of geomythology. The legend is part of the splendid collection made by Ella E. Clark entitled *Indian Legends of the Pacific Northwest*, but the tribal origin of the tale is not given. This story involves the giant Indians we have already discussed in a previous chapter, so we know that the time period, if this story holds up as a historical memory, is very early. We will take portions of Clark's narrative which deal with the geomythological points and then check Vitaliano's interpretation.

> Years and years ago, the mountain peak south of Big River was so high that when the sun shone on its south side a shadow stretched north for a day's journey. Inside the mountain, evil spirits had their

lodges. Sometimes the evil spirits became so angry that they threw out fire and smoke and streams of hot rocks. Rivers of liquid rock ran toward the sea, killing all growing things and forcing the Indians to move far away.

The chief did battle with the evil spirits by throwing rocks down into a crater on the mountain. The battle continued for many days until:

The rivers were choked, the forest and the grass had disappeared, the animals and the people had fled.

The chief knew he had failed to protect the land and sank down upon the ground in exhaustion and discouragement and was soon covered by the lava flow.

When the earth cooled and the grass grew again, they [the people] returned to their country. In time there was plenty of food once more. But the children, starved and weak for so long, never became as tall and strong as their parents and grandparents.[17]

It is said that the chief's face can be seen on the northern face of the mountain.

According to the story, the shadow of the mountain was so great that it cast a shadow that extended a day's walk to the north. The present-day Mount Hood does not cast such a shadow, so this element of the story may also testify to much earlier times than we can anticipate. Vitaliano writes that "although there is no historical record of activity of Mount Hood, the geologic evidence suggests that it may have erupted as recently as a century ago."[18] We should not, at least for the sake of investigating the tradition, assume that the eruption had to have been in recent historical time. It may well have been a very long time ago.

The key to interpreting this legend, it seems to me, is in the casual mention of the *size* of things. The Indians are large, Mount Hood casts a long shadow, and the tunnel under the Cascades is a tunnel, not a bridge. The Indians are reporting accurate facts in their story, but modern interpreters, without telling us what limits they are putting on the story, narrow the possible interpretations to the modern time period and thereby lose the essence of the information which the story contains. No present formations on either side of the river indicate a bridge, but such evidence could easily have been destroyed completely by the gigantic floods that once scoured the Columbia River valley. Almost certainly this legend cannot be referring to an eruption within historic times, since it would take a long time to restore the land and entice the people to come back near Mount Hood to live.

MOUNT RAINIER

Vitaliano cited a Nisqually legend in which Mount Rainier moved from the Olympic Peninsula to the east side of Puget Sound because the mountains on the Olympic Peninsula were growing too big and too fast and crowding Rainier out. On the east side of the sound, Rainier became a monster who devoured everything that came near and finally the Changer in the shape of the Fox subdued her, and she burst a blood vessel and died. Vitaliano noted that there have been some recent lava flows but suggested that since "a volcanic mud flow once poured forty-five miles down the White River valley to the lowlands west of Tacoma, and there spread out in a lobe twenty miles long and three to ten miles wide. ... It is just possible that the 'rivers of blood' are the memory of that event."[19] The mudslide is dated at approximately 5,000 years ago.

Moving Mount Rainier from the Olympic Peninsula to its present location is a highly unlikely geological event. Yet four different tribes of that region repeat the story with but few conflicts in the narrative. I presented Vitaliano's interpretation in an honors seminar in Oklahoma a couple of years ago, expecting the students to accept my denial of the historicity of the event.

In my view this tradition would require the Puget Sound to be located originally north and perhaps east of Mount Rainier to make it appear as if the mountain was on the western side of the sound. We would then need a major earthquake, followed perhaps by a volcanic eruption, or significant mudslide, to bring the waters back around to the western side of the mountain, creating the present configuration of Puget Sound and making it appear that Rainier was now on the eastern side of the sound. In addition, either immediately prior to the event or as part of it, the Olympic Mountains would have to be raised significantly, thus providing the motivation for Rainier to move.

A student in my seminar skipped lunch, went to the library, used his computer retrieval skills, and presented me with a number of articles on the geological instability of the Seattle area. I did not realize until then that a veritable industry had arisen among geologists attempting to pinpoint the possible earthquakes that had occurred in the Seattle area as the so-called Juan de Fuca plate had been encountering the North American plate over long periods of time. This line of research has only arisen since 1987 and, while an increasing number of scholars are working on the subject, it is too early to begin to devise a chronology.

Some of the geological articles bemoaned the absence of any Indian legends describing local seismic events, and it seems obvious that these geologists simply did not know the Indian literature. We can only suggest the scenario that the Nisqually tradition recounts, and predict that the Rainier event was very early in the history of the Pacific Northwest because the reports to date have suggested a lowering of the land, not an elevation.

MOUNT MAZAMA—CRATER LAKE

If we can move to the southern border of Oregon and then move back north, we can discuss the two volcanoes which seem to have the best correlation between Indian traditions and geological data: Crater Lake and the Three Sisters of central Oregon. Geologists call our present-day Crater Lake "Mount Mazama" and have done some basic work in determining when and how it exploded to leave one of America's most unique natural features. A vast scientific literature exists on this crater, but it is written in such technical geological language that the layperson has great difficulty in determining what happened. I will paraphrase an account published in the *National Geographic,* based on the research of Dr. Howel Williams of the University of California at Berkeley, which is the only readable and intelligible source I could find.[20]

Mazama's eruption apparently begins with violent explosions that put a great deal of dust into the air, turning day into night. It calms down for a while (although how geologists can determine this particular part of the sequence escapes me) and then produces a tremendous cloud of steam/dust/ashes. This immensely heated cloud then rushes down the sides of the mountain picking up speed as it moves, basically flattening everything in its path. Lava flows accompany this cloud, although following it somewhat, and literally covers the mountain slopes in several directions, moving in one instance some thirty-five miles. The volcano basically hollowed itself out by producing the lava avalanches and releasing lava through great fissures. Suddenly the peak, which had at one time been at least a mile higher than the present elevation of Crater Lake, collapsed almost straight down into the caldera, producing the configuration we know today. Two important aspects seem to characterize this tale: the heated avalanches and the collapse of the peak into the belly of the mountain.

Ella E. Clark includes a Klamath story about the mountain which parallels the geological scenario quite closely and is worth highlighting. The tradition was related to a nineteen-year-old soldier stationed at Fort Klamath in 1865, some time before scientists would even have seen the lake, much less had time to speculate on its origins. The old man who related the story said it had been passed down from generation to generation, and the soldier asked several other old Klamaths and got basically the same scenario. It happened, according to these elders, "a long time ago, so long that you cannot count it, the white man ran wild in the woods and my people lived in rockbuilt houses. In that time, long ago, before the stars fell. ..."[21] We are talking here about the remote past, a time prior to some major astronomical disturbance that was also remembered.

Personalizing nature, the Klamaths described Mazama and its twin in northern California, Mount Shasta, as having spirits who lived within them, the peaks having an "opening which led to a lower world through which the spirits passed"—indicating most probably that the people had inspected the mountains on occasion and could see the inside of the peaks. The Klamaths knew when the mountain was active because "when he [the spirit] came up from his lodge below, his tall form towered above the snow-capped peaks"[22]—in other words, a cloud of some kind, impressive in its size, was seen.

To cut to the plot, the spirit of the Below-World loved the chief's daughter and demanded she marry him. This amorous overture was denied and the rejection did not sit well with the spirit, so he threatened to destroy the people. "Raging and thundering, he rushed up through the opening and stood upon the top of his mountain." Here we have a cloud and in reality a very angry cloud.

The spirit of Mount Shasta now intervened, conceived by the Klamaths as the chief of the Above-World. A cloud of some magnitude now formed on Shasta, suggesting that it was also erupting, although the actual story seems to indicate a cloud formation of some kind moved down from the sky onto the volcano. The two mountains began some kind of combat.

> Red-hot rocks as large as the hills hurtled through the skies. Burning ashes fell like rain. The Chief of the Below World spewed fire from his mouth. Like an ocean of flame it devoured the forests on the mountains and in the valleys. On and on the Curse of Fire swept until it reached the homes of the people. Fleeing in terror before it, the people found refuge in the waters of Klamath Lake.[23]

I have chosen this description of the avalanche because the Klamath description here fits precisely with the geological version.

The Klamaths then decide that someone should be sacrificed in order to bring calm out of chaos, and two medicine men climb the mountain and jump into the caldera. "Once more the mountains shook. This time the Chief of the Below-World was driven into his home, and the top of the mountain fell upon him. When the morning sun arose, the high mountain was gone." Then, according to the Klamaths, rain fell. "For many years, rain fell in torrents and filled the great hole that was made when the mountain fell. ..."[24]

The only difference I can discern between the geological explanation and the Klamath tradition is that there is a waiting period between the avalanche and the collapse of the mountain. At least enough time existed for the Klamaths to regroup themselves and determine that they needed to make a sacrifice, and during this interlude the mountain cooled sufficiently to allow the two medicine men to climb the mountain and jump into whatever opening existed, to sacrifice themselves. The difference is not material. Indeed, it would make sense to assume that as the volcano cooled, rocks around the edge pulled away from each other, bringing about the final collapse.

But did the Klamaths actually see this volcano erupt? The date of the explosion is estimated at 6,500 years ago, which would place these people at this particular location, as an identifiable group, for a longer period than any other group or nation of people that we know. Sandals and other evidence of human activity in the Crater Lake area have been found beneath the ash layers of this explosion, indicating that *some* humans were eyewitnesses to the event. They can only be the Klamaths in my mind.

What, then, do we make of the reference to a time when white men lived in the wild and the Klamaths lived in rock houses? I personally don't know. Some Indian tribes in the Puget Sound area are considerably lighter in complexion than the Klamaths and the tradition may be referring to them. Minimally, some of the specific points preserved in the legend may be fruitful avenues for future research. Is the "stars fell" reference casually describing a massive meteor shower or a more catastrophic event through which the Klamaths had lived? Scientists would perhaps demand that we discard extraneous information and tie down the tale. It is better to leave a few strings dangling against the day when we are given more information and can add to the story and extend its meaning.

We must, however, ask: If the Klamath account is *not* an eyewitness account, then how did the elder Klamaths come up with a sophisticated version of this event long before 1865, when geology was an infant discipline and was not even ready to begin complex analyses of the sequence of ancient volcanic explosions? How could they have known that Crater Lake is one of the best examples of the top of a volcano collapsing directly downward into the caldera? Most volcanoes apparently blow the top of their peaks or, as in the recent case of Mount Saint Helens, blow out a side as well. As we move on to discuss Mount Multnomah, the question of Indian knowledge of the sequence of geological events becomes even more intriguing.

MOUNT MULTNOMAH— THREE SISTERS

In central Oregon, somewhat east of Eugene and Springfield, is a famous location known as the Three Sisters, plainly visible from most of central, western Oregon. Geologists had visited the site since 1854–1855 when Professor J. S. Newberry of Columbia University examined the area and described it in "Report on the Exploration for a Railroad from the Mississippi to the Pacific Ocean," one of the many surveys done of the western lands in those days. No one, however, suspected the real dimensions of volcanic activity until the summer of 1924 when Edwin T. Hodge of the University of Oregon did extensive fieldwork there.

After examining the Three Sisters and surrounding lava flows thoroughly, Hodge concluded that the then-existing volcanic cones were located within or were a part of the caldera rim of an immense ancient volcano, and he called it Mount Multnomah, using the Indian name of the site. Hodge issued a report a year later, and his reasons for concluding that the location was a former gigantic volcano are worth noting:

... that the Three Sisters mountains rest upon the worn remnants of Oregon's greatest Prehistoric mountain, Mount Multnomah; that this mountain once rose approximately a mile in height above the present snowclad tops of the Three Sisters; that the top of this enormous mountain was lost by a gigantic explosion which left one of the largest calderas in the world; and that the Three Sisters and most of the adjacent peaks have acquired their present form as the result of later volcanic and glacial activity.[25]

In the decades since Hodge's investigation, numerous geologists have verified most of his basic findings and the geological literature is filled with technical papers describing various incidents in the history of this volcanic location. Hodge's summary of the history of the mountain will assist us in looking at the Indian tradition concerning this mountain:

Oregon's greatest mountain was born in Stage "II," or the Oligocene, when an eruption started along the Cascade fault. ... Beginning in the middle Miocene and continuing into the late Miocene, Stage "III," an enormous flood of basic lava poured out. As a result of this intense volcanic activity Mount Multnomah was built into a gigantic cone over 15,000 feet high. ... At the close of the Miocene the entire top of this mountain either collapsed or was blown off. ... Since practically all of the world's great calderas have been due to decapitation by explosion, we may conclude that Mount Multnomah lost its top by such a catastrophe.[26]

The Miocene can be estimated as between 25 million and 27 million years ago, making the mountain quite ancient. There is no doubt, however, that the mountain was once the highest and most impressive of all the volcanoes and mountains in the Oregon part of the Cascade chain. So what do the Indian traditions say about this location?

Again we turn to Ella Clark's collection and find an account so ordinary and commonplace that we wonder why it is included in the book:

Klah Klahnee, the Three Sisters, was once the biggest and highest mountain of all; it could be seen for many miles. One time the earth shook for days, and the mountain boiled inside. It boiled over, and hot rocks came out of the top of it. Flames and smoke rose high in the air. Red-hot stones were thrown out in every direction. Many villages and many Indians were buried by the rocks. When the mountain became quiet again, most of it was gone. Only three points were left.[27]

This tradition comes from the Warm Springs Reservation near Bend, Oregon, not terribly far from the Three Sisters peaks. It has no "supernatural" aspect to it and is simply an account of an eruption of a large mountain with remnants of its former size now seen in the Three Sisters.

Matching the Indian tradition and Hodge's geological account raises certain questions. How did the Indians know that the Three Sisters represented the remnants of old Mount Multnomah unless they had lived at a time when Multnomah was obviously the largest peak in the Oregon Cascades? Here we have an intriguing conflict. According to Hodge, Multnomah reached its highest elevation during the late Miocene and exploded at the end of the Miocene, which ended approximately 25 million to 27 million years ago. If anyone entered the area after the explosion, it would have been difficult to identify the ruins as representing the largest volcano in the northern Cascade chain. Indeed, either Mount Rainier or Mount Hood would immediately be seen as the largest mountain in the chain.

Are we to believe that the Indians were eyewitnesses of the event, as they seem to have been? It seems to me that either we credit the Warm Springs people with residency of 25 million years, or we credit them with a geological knowledge in the 1850s far in excess of anything achieved by white scientists until 1925. If the Warm Springs Indians crossed the Bering Strait around 12,000 years ago, they had to have come to the Oregon area *after* Mount Multnomah had exploded and had long since begun to erode. They could not have known that this particular mountain was once the highest of all the Cascade peaks.

We have to remember that Indians did not wander the Cascades trying to explain the origins of mountains and their possible relationships. Indeed, aside from occasional hunting or religious vision quests, most of the Indians remained on the lower lands because there was a fear of mountain spirits and a sense of religious awe regarding mountains. And hunting and fishing are not exciting activities above the tree line. Other legends in Clark's book make clear that many tribes did not want whites to climb some of the mountains for fear the spirits would take offense.

We have another alternative in explaining the conflicting interpretations: the geological time scale is wrong. The mountain was once the largest peak in the Oregon Cascades but the eruption was well within the memory and experience of man. Curiously, Hodge himself hints at this solution because he notes: "The most striking peculiarity of the Three Sisters region is the obvious youth of the many volcanic floods, volcanoes, and cinder fields." He also relates that "these black, scoriaceous, volcanic rocks look so young that many are convinced that they have congealed within historic time. These congealed lavas in total cover seventy-eight square miles and form one of the largest recent igneous floods in the United States."[28]

What is it about scientists that they observe with their own eyes the obvious youth of volcanic rocks and yet, apparently because of doctrinal considerations, reject their own sense perceptions and classify evidence according to a predetermined scheme? It is this stubborn application of abstract orthodoxy to real-life situations that makes science a hilarious farce in many areas of endeavor. The Hodge identification coincides with Derek Ager's constant observation that volcanic rocks in Europe look so fresh they might still be warm. If volcanic rocks look exceedingly fresh and have little if any erosion, at least the erosion one might expect to happen in 25 million years, maybe they are fresh.

Scientists may argue that potassium-argon tests would show the age of selected samples of the Multnomah lava and that the orthodox geologic dating was correct. Potassium-argon testing is measuring the percentage of potassium-40 isotope to argon-40 isotope in a rock. The assumption is that, during the formation of the rock, *no* potassium had yet begun the breakdown process of becoming argon. The measurement determines, from the percentage of each element, how old the rock is, given a constant rate of decay. This kind of reasoning, if you will remember, once enabled Mark Twain to write a little ditty on the Mississippi suggesting that the Mississippi was once so long that it stuck out into the Gulf of Mexico for hundreds of miles and that sometime in the future it would be but a few miles long.

Although most scientists swear by the K-Ar dating system, the tests are notoriously unreliable. Cremo and Thompson point out in *Forbidden Archeology* that, using the potassium-argon method of dating, "... scientists have obtained ages ranging from 160 million to 2.96 billion years for Hawaiian lava flows that occurred in the year 1800."[29] When it comes to providing dates for volcanic rocks, then, it would be better if people just guessed, and if the rocks looked very recent, they should not hesitate to admit it.

SUNSET CRATER

Scholars have disputed Indian stories about very recent volcanic explosions, dismissing the Indian accounts as superstitions according to doctrinal requirements of anthropology. In the October 1932 issue of *Museum Notes*, published by the Museum of Northern Arizona, Harold S. Colton related a story told by the Hopis of a possible eruption of Sunset Crater, which is located about sixty miles southwest of Oraibi. The tradition can be briefly summarized: For four days and nights Hopis saw fire and smoke coming from the Sunset Crater area. Some of the people grew alarmed when it appeared the fire might be coming their way, but nothing was done. Colton dismissed the possibility that the Hopi had witnessed an eruption, and his analysis illustrates the great difficulties Indians have in getting their traditions a respectful hearing.

Although the tradition may be considered to refer to a volcanic eruption, yet it should not be given too much weight. The tradition deals
with the destruction of a pueblo seven miles northwest of Oraibi,
Pivahonkiapi, which, from the pottery, we know existed from about 1100
to 1300, dates much too late for the Sunset eruption. That the eruption
is supposed to have lasted four days has no significance, as all events in
Hopi traditions take place in fours; four being the Hopi ceremonial
number. The tradition, therefore, must be digested with a grain of salt.
It may not refer to anything more than a forest fire.[30]

While Colton tentatively suggested that this tradition might involve the observation of a volcanic eruption, that it lasted four days,
and that anthropologists had already dated an adjacent pueblo
via their pottery classification were for him sufficient reasons to
declare the event simply a misreading of a forest fire.

What if, in fact, the fire *had* lasted four days? How would the
Hopi have expressed this fact in order to gain credibility with white
scholars? Does "four" have anything whatsoever to do with the
observation of a nearby geological event? If a white observer said
the eruption lasted four days, would we believe him? How would
white people like it if other peoples automatically discounted anything that dealt with "three," since the Holy Trinity of Christianity
involves that number? Would that necessarily mean that our three-
day weekends are mythical or holy because of the correlation of
numbers? Would we discount the division of our government into
three branches because "three" happened to correlate with the
Trinity? Does "three strikes, you're out" in baseball, and now apparently in crime, have some kind of theological significance?

It so happens that the Hopi have recently published a book
which deals with their tradition of the Sunset Crater eruption.
They have linked the disturbance with a story of the marital
problems which arose when a Hopi girl married a man who was
a kachina but was betrayed by the envy and jealousy of the
people of his village. The young kachina built a fire but it got

out of hand and burned down into the underground world, sparking the subsequent eruption. The Hopi description of the eruption is worth noting:

> When the people learned of the fire, they looked across to Nuvatukya'ovi. Right inside one of the hills there was a mass of what looked like boiling coals pushing upward. As this was going on, the fire rose sky-high. Some time later, as it reached its peak, it spewed out molten embers that were extremely hot. Bursts of sparks kept shooting into the air just as from a torch someone is running along with. Finally the embers reached up so high that they poured over the rim of the fire pit, thereby enlarging the hill. Eventually the hill sprung leaks at several places around its base. Spewing forth in every direction, the molten embers started running everywhere. Nothing seemed able to block the flow. From all appearances, the flow was heading straight toward the village of Musangnuvi. By now, the embers were only oozing out from underground and flowing outward. The top of the hill had apparently cooled off and was quiet. But the hill had increased in size considerably. Also due to the cooling action there now stood an enormous mountain. In its vicinity several other mounds had formed but they were not as large.[31]

From this vivid description of the mechanics of the eruption it would seem likely that we have here an eyewitness account, whether it is mixed with a romantic story and has a supernatural component or not. Indian traditions involving mountains and volcanoes raise a series of very important questions. Any single account can be subjected to intense analysis and critique, and scholars can say that they do not accept some version of how a volcano erupted. But many of the Indian accounts were put down on paper before any non-Indians were able to visit the locations and apply "scientific" knowledge to the evidence and offer their own scenario.

What happens when the Indian account and the geological account do not vary significantly and many small factual details seem to match? The mathematical odds of several tribes of Indians always guessing correctly about the origin of specific locations and the sequence in which they came into their present form increase substantially when we begin to link up some of the traditions and show correlations that are consistent. When the Indian account includes the mechanics of the event, it should be regarded as a valid eyewitness account.

This knowledge of geologic and climatic events in the North American ancient past preserved by the traditions of the tribes can be a significant source of information for modern science. But it would require that scientists honestly reevaluate much of their dating of strata and abandon orthodox doctrines in instances where common sense dictates otherwise. Fresh-looking lava must be reasonably recent; processes of erosion cannot be suspended, like scientific beliefs, simply for doctrinal purposes.

If the Indian legends demonstrate the presence of people in North America, or even the Western Hemisphere, tens of thousands of years ago—or in the case of Mount Multnomah 25 million years ago—then that discrepancy should alert scientists and they should reexamine their doctrines in light of the conflicting interpretations. The idea that people have only been in the Western Hemisphere for 12,000 years is simply an agreement among scholars who neither think nor read and who have been stuck on a few Clovis and Folsom sites for a generation. I personally cannot believe that any people could remember these geological events for tens of thousands of years. *My conclusion is that these are eyewitness accounts but that the events they describe are well within the past 3,000 years. It is past time that this resistance be ended and a new scenario for the Western Hemisphere be constructed.*

Floods, Lakes, and Earthquakes

VOLCANOES PROVIDE THE EASIEST natural phenomenon to link to tribal traditions because they are not difficult to date in the geological strata, although we have seen that the potassium-argon method of dating leaves much to be desired. Nevertheless, volcanic strata are beneath or above other strata that give us some sense of historical sequence. Matching traditions about floods and the creation of lakes, rivers, and inland seas is somewhat more difficult, since water is an erosive force that can wipe out otherwise useful signs of age.

Written material suggesting an Indian knowledge of lakes and rivers has been radically transformed by people who were interested in oblique topics not directly related to the question of Indian knowledge. Almost all of the stories about lakes bring with them the romantic story of Indian maidens pining for lost or forbidden lovers. Flood stories are almost always linked with the concerns of fundamentalist Christians who believe that Indian accounts of a great flood will provide additional proof of the accuracy of the Old Testament. With their cultural blinders in place, it never occurs to them that the Old Testament may very well provide evidence of the basic accuracy of the Indian story.

Scholars in comparative religion, anthropology, psychology, and folklore usually steer well clear of using flood stories for anything except demonstrating that all societies have these kinds of traditions. The most common superficial explanation is that flood stories respond to a basic human psychological need and are therefore a part of the orientation process that societies devise to enable them to live in this world. But these flood stories almost always have geographical references to mountains where there are high-water marks and to locations where people lived prior to the flood. And they are stories that often provide a historic framework into which other experiences from the past are placed. The obvious question that presents itself is whether these flood stories do speak of a planetary event not so long ago involving significant psychological trauma that caused the event to be remembered.

Surveying the Indian memories of volcanoes and floods, one can immediately see that the short-term duration of volcanic eruptions leads the people to interpret the eruption as the work of a spirit, generally of the mountain, or of a number of spirits, depending upon the scope of the violence—one variant of the Three Sisters suggests that the mountain spirit of Multnomah had three wives who got out of hand; the Crater Lake story involves two mountains and two powerful spirits. Surprisingly, many flood stories include volcanic eruptions as part of the scenario, so that the stories suggest physical disruptions on a substantial geographical scale.

The pervasive nature of the large floods, the wide geographical scope of their damage, and the seemingly complete destruction of the world as people have known it lead many tribes to remember the experience as a general purging of evil in the world. The tribal accounts therefore need to be "demythologized," not in the old Rudolf Bultmannian search for enduring religious truths, but simply to eliminate the idea of crime and punishment so as to allow a concentration on the physical phenomenon of an unusually spectacular and destructive flood event.

Let us begin our examination of Indian flood stories with the traditions of the Indians on the Pacific Northwest coast. These groups are basically seafaring peoples who used the sea as the Plains tribes used the land. With a complex set of triangulation devices, these tribes—Quinaults, Makahs, Clallams, and others—went far out to sea, hunting whales and seals. Therefore, over time, they experienced the terror of the sea as well as enjoyed its more placid benefits. If any groups knew and understood tsunamis and other hazardous ocean activities, these would be the people.

Their flood stories, for the most part, involve tsunami actions of unusual strength and duration. Mount Shasta legends say that the sea came inland and rose until it nearly covered this peak and then finally receded, leaving behind dry land and the marshes of the northern part of the state and southern Oregon. Strangely, no flood stories were collected by Ella Clark from the Oregon tribes, and my suspicion is that these tribes may not have survived the flood, or may not have been affected by it. If there are traditions among these groups regarding a flood coming from the west characterized by rising tides, then they are yet to be recorded.

When we reach Washington State we discover that hardly an Indian group exists that does not have a flood story, almost unanimously involving the sea invading the land. The people's solution, having been forewarned, is to build canoes, sometimes rafts, and attempt to ride out the storm. In a number of stories only the good people in the canoe are saved. The majority of these stories appear to involve the efforts of the Indians to survive by fixing their canoes to the tops of mountains. They then identify landmarks and peculiar geological formations on the mountains as the site where the canoes were tied.

More often, however, this flood separates the different canoes and the tribe is scattered over a vast area before the water ebbs, leaving the people isolated from each other. The Skokomish, for example, scattered so widely that one group traveled far to the east and became the Flatheads, which are today a combination of Salish and Kootenai people. "A long time afterward," as an elder told Ella Clark, "when there was war around where Seattle is now, the Skokomish people were trapped on the bay. They heard strange people talking in the Skokomish language.

When my people spoke to them, they said, 'We are the people who drifted away from here.' That is why the Skokomish and the Flatheads speak the same language."[1] The Quillayutes say:

> For four days the water continued to rise. At last it covered even the tops of the mountains. The boats were carried this way and that way by the wind and the waves. The people could not guide them, for there was no sun and there was no land. Then the water began to go down. For four days it receded. By that time the people were scattered. Some of the canoes landed along the Hoh River. So those people have lived there ever since. Others landed at Chemakum, on the other side of the mountains. They have lived there ever since. Only a few found their way back to the Quillayute river.[2]

This story closely parallels a flood story of the Makahs who live on Cape Flattery. The tradition is cited in the earliest study of this tribe by James Swan, a schoolteacher and doctor for the tribe who was appointed immediately after the Makahs signed the 1855 treaty of Neah Bay with Isaac Stevens. Swan was able to verify parts of this tradition by examining the prairies near Neah Bay to confirm that the sea had indeed invaded the cape in a major happening. Although it resembles the Quillayute tradition, the story is worth repeating:

> A long time ago, but not at a very remote period, the water of the Pacific flowed through what is now the swamp and prairie between Waatch village and Neah Bay, making an island of Cape Flattery. The water suddenly receded, leaving Neah Bay perfectly dry. It was four days reaching its lowest ebb, and then rose again without any waves or breakers, till it had submerged the Cape, and in fact the whole country, excepting the tops of the mountains at Clyoquot. The water on its rise became very warm, and as it came up to the houses, those who had canoes put their effects into them, and floated off with the current which set very strongly to the north. Some drifted one way, some another; and when the waters assumed their accustomed level, a portion of the tribe found themselves beyond Nootka, where their descendants now reside. ...[3]

The Makahs and the Nootkas, therefore, are basically the same people, separated some time in the past by a monstrous tidal wave. The warm water mentioned in this story suggests that some kind of oceanic volcanic disturbance was involved, although there is a chance that an unusually strong version of the Japanese current may have been responsible.

Matching the various Indian descriptions of this tidal wave along the upper Washington coast is not difficult if the locations of the tribes are taken into account. The coastal tribes, such as the Makah, describe a rather steady and persistent rise of the water, enough so that the people could get prepared for the inundation. Some accounts that have been given to me orally indicate that the water first went out to sea, was gone for four days, and then came back in a rush. Since the Swan version is earlier, I suggest that description should take precedence.

When we examine the accounts of tribes in the inland sound area, such as the Clallams, Squamish, and Swinomish, the flood waters are described as if they had developed an increased velocity. It would seem likely that as the tidal wave reached the narrow channels in the Strait of Juan de Fuca, and later as its waters hit the San Juan islands, the velocity of the wave would have increased significantly, producing a different physical description altogether. I would argue that we are talking about one event seen and experienced from several perspectives. But what kind of tsunami would create a wave so tremendous that it would submerge the peaks on the Olympic Peninsula and continue on to reach the snow line at Mount Baker and other peaks far inland?

The steady swell of the rising waters on the beach at Cape Flattery seems to me to raise doubts about this event, providing evidence of a planetary disaster such as the one required to create the Ice Age. Any extraterrestrial source approaching our planet would create havoc of unimagined dimensions and the waves would be far more spectacular. The best we can suggest from these accounts is a very major kind of activity in the Pacific Ocean basin.

These flood stories that involve tidal waves must be distinguished from flood stories that have rain as their primary source of water. Tribes all over the country have flood stories that feature incessant rain as the source of the disorder, and these stories, if any Indian stories, may have some relationship to Noah's flood. Otis Halfmoon, a Nez Perce elder, told Ella Clark that "it rained for a long, long time. The valleys were filled with water, and the animals lived on the tops of the hills. Some of the animals were saved, but the big animals perished. That is why people have found the bones of big animals along the Salmon River and big hip bones near Lewiston."[4] This tradition fits comfortably with the argument earlier in the book concerning the demise of a major portion of the megafauna at the time of the large ice/water dump.

The rain scenario is also found on the Pacific coast, although having lived in that area, I cannot help but imagine that, if these Indians remembered a specific rain distinguishable from all others, they were talking about a pretty big event. My memories of the region are that it began raining January 1 and continued until December 31 with barely a break in June to glimpse the sun briefly. The Skokomish relate that the Great Spirit was displeased with the evil in the world and after having secluded the good people and animals, "... he caused a heavy rain to fall. It rained and rained and rained for many days and many nights. All the earth was under water. The water rose higher and higher on the sides of Takhoma [Mount Rainier]."[5] The water did not subside until it had reached the snow line, so we may be talking about a substantial amount of rain. But we must remember that Mount Rainier was also actively shifting its location, or having the location changed for it, so that it may not have been anywhere near as tall then as it is today.

The Quillayute Indians who live on the Olympic Peninsula have a tradition which fits perfectly with the icy comet scenario. They experienced a storm of vicious intensity and prolonged duration. According to their tradition: "For days and days great storms blew. Rain and hail and then sleet and snow came down upon the land.

The hailstones were so large that many people were killed. The other Quillayute were driven from their coast villages to the great prairie, which was the highest part of their land."[6] The storm lasted so long that the people grew thin and weak from hunger, and it was so intense that the men could not go out on the sea to fish. Since these people always had a ready supply of food in this area, the long duration of the storm seems to suggest that we are dealing with a major climatic event. I would date this storm of rain, hail, sleet, and snow as preceding the tidal flood which divided this tribe into smaller scattered groups.

The lake and river flood stories are much easier to date, and demonstrate, in a manner reminiscent of the volcano stories, the possible historical basis of the tribal traditions. The most prominent tradition involves the Nez Perce, the Yakimas, and the Spokanes and involves the eastern Washington scablands flood which scoured out the Columbia Valley when giant Pleistocene Lake Missoula broke and emptied. Theoretically, this outburst was the largest such event in planetary history, although a Russian geologist has recently claimed that an even greater flood of approximately the same geological time period occurred in Siberia. Identifying two monstrous floods in these northern regions means that the glaciers must have melted at an unreasonably rapid rate, thereby lending some credibility and support to Donald Patten's scenario.

Identifying the area and nature of the Washington scablands event (or events) is a story in itself. J. Harlan Bretz, a local scholar intensely interested in the scablands, began to investigate this landscape in the 1930s and quickly concluded that it was the scene of an impossibly large flood. The scablands is a vast area southwest of Spokane along the Columbia River, featuring dry river channels, massive waterfalls, and thousands of sand bars with some major dry pools and sinkholes, indicating that the landscape was carved by an almost unbelievable amount of water rushing at an enormous speed. The scablands region, according to Bretz, had suffered possibly two major floods beginning around 10,000 B.C.

The water had run from the Spokane area across to the Grand Coulee, then south down the main channel of the Columbia, with backwash areas east and west of the Columbia channel, then down the present Columbia, backwashed into the Willamette Valley, and finally pushed through to the Pacific Ocean. Even to imagine an event of this magnitude in the 1930s was to incur the wrath and disdain of everyone in the scientific community, for doctrine had decreed that no catastrophic events had ever occurred or would ever occur.

The respected giants of geological orthodoxy bitterly attacked Bretz's ideas, and for most of his life he was ridiculed and ostracized by his profession for suggesting the possibility of a catastrophe of this scale. Bretz continued to advocate the theory and to publish additional papers elaborating on the idea. When he was in his early nineties a group of geologists finally gave the idea some credence, came to eastern Washington, and walked some of the terrain as Bretz had asked them to do for decades. He got an apologetic telegram from the group at his home following the tour, declaring that the geologists were now all catastrophists.

With the Bretz theory established, the next generation of geologists felt safe to speculate on the nature of the disaster. And many younger scholars, now able to work in catastrophic theory because it had become orthodoxy, began publishing papers on their version of the scablands geological history. At the present time, as many as forty different floods have been suggested, demonstrating that scholars can frequently get out of hand, given a novel idea to contemplate.

John Allen, Marjorie Burns, and Samuel Sargent, friends and colleagues of Bretz, summarized the present state of scablands flood literature in a delightful book, *Cataclysms on the Columbia*, in which they discuss the trials of Bretz in getting the idea a hearing, his eventual triumph, and the consensus scientific beliefs about the events. Their preferred date for scablands activities is between 15,000 and 12,800 years ago, and they seem inclined to accept a much larger number of floods, indicating a gradual melting of the ice sheets, taking away much violence from the initial outburst and distributing it over 2,000 years and the forty floodings.

The present scenario conceives of the Cordilleran ice sheet with a lobe penetrating the Purcell Trench in British Columbia and holding back billions of acre-feet of melted glacial water. This lobe then begins "advancing and retreating" between Pend Oreille Lake and the Clark Fork River. "Each time it advanced up the Clark Fork several miles it formed an ice dam as much as 2,500 feet high across the valley impounding the waters behind the dam to form a great lake up to 2,000 feet deep, covering 3,000 square miles, and extending for 200 miles to the east in the intermontane valleys within the Rocky Mountains."[7]

The authors designate this area as "Lake Missoula" and they estimate it was nearly one-half the volume of our present Lake Michigan. Allen, Burns, and Sargent paint an idyllic picture of the plains of eastern Washington just prior to the bursting of the Lake Missoula ice dam:

> Wooly mammoth and mastodon, longhorn bison, camel, caribou, and musk ox roamed the plains. Close behind, and preying on these herbivores, came the hunters, wolves in packs or lone saber-toothed tigers hunting by stealth. Giant condors soared across the skies, in search of carrion left by the hunters; and along the tributaries to the main rivers, giant beavers built dams, while shaggy, short-faced bears vied with early humans for the abundant salmon that swarmed up the Columbia and its side streams to spawn in the shallows.[8]

With this peaceful scene established we must at least recognize that the scablands floods, not mischievous Paleo-Indians, were responsible for the loss of megafauna in this region. But were there any Indians present at the time of the flood?

Bretz did not believe that Indians were around during the first flooding, and the only evidence we have that he believed that Indians were around for *any* of the flooding is simply this comment by his friends and disciples: "The only direct evidence in favor of

Early Arrivers in the region of the Bretz floods," Allen, Burns, and Sargent suggest, "is a campsite with charred bones and stone artifacts buried under pre-flood deposits at the site of the Dalles Dam and a single stone artifact recovered from a Bretz Flood gravel bar at the mouth of the John Day River."[9]

The problem here is that if you have Paleo-Indians coming across the Bering Strait 12,000 years ago, you have to move them very quickly across the Alaskan mountains, down the so-called ice corridor to approximately Montana, and have them do an abrupt right-hand turn south of the glacial sheet and head directly across Idaho to the central Washington area. The odds of this kind of travel are minimal to impossible. Unfortunately, Allen, Burns, and Sargent simply lapse into familiar anthropological doctrine at this point, stating:

> In the Old World there were human and human-like remains going back in older succession through the essentially human Cro-Magnon, the less human Neanderthal Man, the even less human *Homo Erectus*. ... [10]

And so forth, repeating the outmoded sequence of alleged human evolution. And they follow the party line in suggesting that "... the transit of the two continents may have taken only a few hundred years—along with the extinction of the elephants, mastodons, camels, and most of the other large game animals of the New World."[11] At the rate "scientists" are reducing the time required to have extinguished the megafauna, we will soon be told that a single Paleo-Indian glared across the Bering Strait and thirty-one species of megafauna fell over dead.

Derek Ager suggested that Amerindians had probably arrived in the area sometime before the flood because the "Marmes Rockshelter, near the spectacular Pelouse Falls in Washington, has yielded human bones and human artifacts. Probably many of these early people perished in the flood."[12] It is not difficult to see that admitting the presence of humans prior to the flood will involve serious scholarly disputes about the age of any sites located within the area which are beneath flood deposits. These sites would probably not be acceptable to the anthropologists and archaeologists who are wedded to the Bering Strait theory. Allen, Burns, and Sargent, however, are all optimistic about finding evidence of Indians in the future in this area.

At any rate, let us deal with the first flood and its intensity and then see what the Indian traditions relate. Allen, Burns, and Sargent describe the event in unforgettable language:

> The mass of compressed air—impelled by the towering head of onrushing water—hit first. We know about winds building over time into hurricane strength, but how can we imagine a torrent of air exploding into existence, driven by a wall of water hundreds of feet high and moving at 50 miles per hour? ... think of these shock waves as emanating from the face of one plunging, thundering, rolling-water-wall, moving at speeds only slightly less than that of a car on a freeway, a swollen, surging mass composed of ice, rock, mud, and water, and standing over 500 feet tall for the first mile of its discharge, and again this tall where the waters were compressed and funneled by constricting features such as the Wallula Gap and the Columbia River Gorge.[13]

The flood, then, if one survived, was really something to tell one's grandchildren about. Allen, Burns, and Sargent estimate that "the maximum flow was more than 9.5 cubic miles of water per hour, which could have drained the lake in two days. More probably it slowed and lasted for a week or more."[14]

Many tribes have traditions concerning a great inland flood, and a significant number of people believe it was the Spokane scablands disaster. We should perhaps begin the Indian accounts with a tradition of the Spokane tribe. The Spokane Indians lived in the area north and northwest of the present city of Spokane, Washington, and I have located at least two versions of their flood tradition that relate to this particular flood. Ella Clark's version is found in her *Indian Legends of the Pacific Northwest*, and Deward Walker's much shorter account comes from Alex Sherwood, who for many years was chairman of the tribe. We will look at Walker's version first, and then Clark's account will be used to expand our scope of discussion.

Many years ago the Spokane area was a large lake which took several days to cross. There were many villages around the lake and on the various islands of the lake. ... One bright morning tragedy struck. The earth started rumbling and shaking. The startled Indians fled before the waters as huge waves pitched into the air, overturning boats and engulfing villages. The game drowned as they tried to escape to higher ground. Many died of starvation and thirst. Mount Spokane, the Little Mountain that grew big overnight, gathered some to her care. Then the earth sucked the whole lake into the "World Below." The lake was gone.[15]

Mount Spokane, it appears, had originally been an island in this gigantic lake. As the water drained to a lower level it seemed to the Spokanes observing the event as if the mountain was in the process of growing out of the ground.

Ella Clark's version of the tradition was told by Chief Lot of the Spokanes to an army officer on the Spokane Reservation around 1890, a few years before J. Harlan Bretz was born, thus eliminating the idea that the Spokanes might have heard it from him. This version contains a bit more information, which helps illuminate the scope of the story for us. Chief Lot reported that the original lake was many days' travel long, indicating that it must have extended well into Idaho and perhaps even into Montana. The disaster began with an earthquake that agitated the waters of the lake considerably. After the water rose to become an awesome set of waves beating relentlessly on the shores:

Then the sun was blotted out, and darkness covered the land and the water. Terrified, the people ran to the hills to get away from the pounding water. For two days the earth rumbled and quaked. Then a rain of ashes began to fall. It fell for weeks. ... At last the ashes stopped falling, the waters of the lake became quiet, and the Indians came down from the hills. But soon the lake began to disappear. Dry land rose where the water had been.[16]

The longer version of the story suggests that an earthquake, or quite possibly volcanic rumblings indicating a rise of the magma to potentially dangerous levels, was the first sign of danger. The possibility exists that a tremendous volcanic eruption in some other part of the Pacific Northwest is associated with this emptying of the great Pleistocene lake, since in Chief Lot's version, ashes fall for weeks and the scene calms down before the lake begins draining. It is not difficult to see that Alex Sherwood's version picks up precisely at the point where Chief Lot's concludes, the difference being one of emphasis by the two Indians.

In Clark's version the people eventually follow the river to the west until they find a waterfall, perhaps the Grand Coulee or maybe one of the other fossil waterfalls which we see today. When they see salmon coming upstream in the new river channel, they realize the catastrophe has passed and they settle along the new river. It is difficult to imagine the magnitude of the flood in order to hypothesize when the river system would again be ready for salmon migrations. Certainly there would be many years of turbulent mud deposits and total ruin of spawning habitats. But the salmon, safe in the Pacific Ocean as part of their life cycle, would not have been destroyed and would have returned, as best they could, to areas which represented or resembled their original home.

Chief Lot's account contains sufficient data so that we can question the orthodox interpretation of the glacier "advancing and retreating," creating a series of ice dams, and engaging in sporadic floods over a long period of time. By collapsing the time frame in which the scablands floods occur to this one major event lasting several weeks, we can link the Indian traditions to the scablands floods and not become involved with the prolonged period of time (2,000 years and perhaps some seventy floods) required by uniformitarian interpretations.[17]

Many arguments can be made on behalf of the Spokane version of the flood. First, if these floods were periodic, Mount Spokane would not have appeared to have suddenly grown out of the ground. Rather, the Indian tradition would involve a story in which the spirit of the mountain spent his (or her) time traveling back and forth between our world and the "World Below," struggling with a monster that was inhibiting his growth. The fact that Mount Spokane did not become the subject of continuing stories about rising and falling argues in favor of one large flood and the permanent draining of the lake.

Second, the Spokane tale clearly marks this event as being triggered by severe volcanic eruptions and provides a time frame, the several weeks of falling ashes, before the dam breaks and the lake drains. If the lake was as large as the Indian accounts suggest, that it took many days to travel the length of it, then the waters were impounded by solid rock and not a tenuous ice lobe that itself might have been melting in rapid fashion. So it would take a major earthquake and/or volcanic eruption to break through the solid rock walls of the lake, or tilt the land significantly, to open a passageway to the west and initiate the massive flood.

Third, and simply common sense, if there were as many as forty floods we should then have forty advances of the Clark Fork lobe of the glacier and thirty-nine retreats. In order to have this kind of activity we need periodically to warm and cool thousands of square miles of glacier located due north of Spokane Lake. That would mean, over a period of some 2,000 years, a warming and cooling every thirty years—a feat possible only in the minds of scholars. But this advancing and retreating scenario is ridiculous even in scientific terms. C. Warren Hunt examined the proposition of having an ice lobe act as a ditch gauge, releasing and impounding water to provide for the many floods that modern geologists now want. Pointing out that modern engineering techniques require considerable bedrock to secure the footings of dams to fill a space as small as five hundred feet, he implied that it was ridiculous "to suggest that chance emplacement of glacial ice might have dammed Clark Fork across a 7-mile ... span lacking in intermediate abutments and then retained water at four times the pressure of modern engineered, concrete dams!"[18]

Again we are dealing with casual, almost flippant, scholarly speculation passed off as reasoned discourse when we accept the idea that a multitude of floods occurred and were caused by glacial lobes sporadically creating dams, which strangely were much stronger than engineered modern dams.

We have little written testimony of the traditions of the Indians living downstream of the Spokane area on the Columbia River. The vast majority of them must have been obliterated with the first major flood. Ella Clark has a very short entry regarding the flood from the perspective of the downstream people: "Steptoe Butte stood above the waters at the time of the great flood, and many Indians there were saved from drowning. Below the top, there used to be a water line, the mark of where the water had once been."[19]

The Nez Perce called Steptoe "Ya-mas-tas" and said that they had survived the great flood on top of that mountain. It was subsequently called the "Holy Mountain." Across the Columbia to the west, the Yakimas said that they had survived the flood in a big canoe, which eventually lodged on Toppenish Mountain. We cannot be certain that the Nez Perce flood was the same as the scablands flood, because there was a large flood on the Snake River moving westward at one time, and there is not sufficient identification of the event to distinguish a particular flood. Since Steptoe Butte is somewhat removed from the Snake River, it would seem reasonable that the Nez Perce flood story is the scablands event and that they were talking about one of the backwater surges that were characteristic of that event.

Pleistocene Lake Missoula was a tremendously large body of water, and being located high in the Rocky Mountains meant that its arms extended quite a distance into the various valleys that mark the western side of this mountain chain. We could expect, therefore, some traditions that would describe some of the minor events in the life of this gigantic lake. The Flathead people (Salish and Kootenai) who live in the Flathead valley northwest of Missoula, Montana, have preserved a tradition that describes a powerful flood that created Flathead Lake. "The great water first came to the valley where Flathead Lake has remained to this day. The flood grew bigger and bigger, spreading over all the lower lands. Most of the people were drowned in the valleys, but others fled to the high-est mountains. At last all the land was covered except the solitary peak where a few Indians had gone for refuge."[20]

The people, rapidly being consumed by the waters, prevailed upon their chief to halt the flood and he futilely shot two arrows at the rising water. They simply floated away. His third arrow stuck in the ground but the water came up to the third feather before it crested. Then "gradually the water went down, the tops of the mountains appeared, then the hills, then the valleys. The water that remained became Flathead Lake."[21] The tradition has a ring of authenticity to it. It seems to me that two possibilities exist: (1) the Flathead flood is a local flood brought about by the *formation* of Lake Missoula; (2) a major landslide or ice-slide moves into an existing Lake Missoula and pushes the water south; the water then sloshes backward to the north again, draining Flathead Valley and leaving a certain amount of water and thus creating the lake.

Some interesting possibilities exist with the Flathead traditions because they have other stories about how the lake once drained to the north, opposite its present outlet. The story about the south-ern outlet involves a giant beaver, which may argue that it was earlier in time. A flood coming from the north, however, would assist in creating a southern outlet, deepening whatever then ex-isted considerably and making the lake permanently drain to the south toward present-day Missoula.

Since Pleistocene Lake Missoula was supposed to have covered the present-day city of the same name, it would seem likely that the first flood story must take precedence chronologically and the story describing the southern drainage with the giant beaver must be the later event. We then have evidence that in the Missoula area the giant beavers survived the scablands flood and perished at a later date, a fate which does not correspond with stories of other tribes that all the giant animals were destroyed in the flood. So the subject is primed and ready for more precise scholars to examine.

The scablands flood presumably moved into and scoured out the valley of the Columbia River, and the assumption of the Bretz school of geologists is that the Columbia River predated the flood. A Yakima story entitled "How the Coyote Made the Indian Tribes" sheds some interesting light on the origin of the river. A giant beaver inhabited Lake Cle Elum on the eastern side of the Cascades. His name was Wishpoosh and he abused the people so that Coyote decided to help them.

Coyote and Wishpoosh got into a fight in Lake Cle Elum and caused an earthquake which made a large hole in the lake, and it began to drain. Wrestling with each other and refusing to give in, Coyote and Wishpoosh rolled down the eastern slope of the Cascades to Kittitas valley, where the waters made a great lake. The combat continued on, Coyote and Wishpoosh struggling with the waters rushing behind in their wake. They cut the channel for the Yakima River, created a second lake, and tore through Union Gap. The waters overflowed this path and formed another lake in the Walla Walla country. The fight then took an abrupt turn to the left and the Oregon-Washington border channel of the Columbia was made to the Pacific Ocean.[22]

The Yakima story is echoed in several other tribal traditions where only part of the sequence is mentioned; the Colville, Sanpoil, and Okanagon tribes all repeat parts of this story. We can, of course, minimally claim that the tradition verifies the existence of the giant beaver. But of more interest is the identification of the river channels and lakes that are created during this conflict. Some minor versions of the scablands flood stories emphasize the existence of lakes in the lower eastern Washington area that are destroyed as a result of the tidal wave from the north. There lakes could well have been produced in the tidal wave issuing from Cle Elum. Again we have a knowledge of ancient geological features that modern Indians would not or could not have known unless their traditions remembered them.

Although the stories about lakes and floods in other parts of the country are sparse, there is one tradition that bears our examination. Ella Clark includes a lake story from the Shoshone tradition that represents this puzzling aspect of the Indian accounts. It begins: "Long, long ago, the Big Horn Basin was a great sea." Fish and animals, although not megafauna, lived on the shores of this sea and were too smart to be caught by the Indians, so the people were thin and hungry. "Suddenly the waters of the great inland sea began to lower. Down they went, the Indians following, until the water was so low that fish were piled on top of each other." The people had their fill and even preserved some of the fish. Then "at last the great sea disappeared completely, but there was left behind a river that roared through a crack in the mountains."[23] A straightforward narrative with no supernaturals present.

We seem to have here the creation or origin of the Big Horn River. Although the full story has reference to prayers of the people who were starving and needed to catch fish, the real story line is simply that the lake emptied into a new river and the fish were caught. Now the question arises how the Shoshones knew that the Big Horn basin was once an inland sea. We can suggest that they found seashells or that a visible waterline existed hinting at the original state of the basin. But if such a waterline existed, wouldn't we expect that it would be cited as evidence of the great flood and not be identified as a seashore? I suspect that the story provides evidence that Shoshone occupancy of this region goes back a long, long time.

The Pacific Northwest is the perfect area to match Indian traditions and geological knowledge because of its many unique geological features. Where else can we get rivers, scablands, volcanoes, lakes, and floods within a restricted area and all in such close proximity that some stories describe the relationships between volcanoes and rivers, lakes and earthquakes? It is the interlocking of geological phenomena that makes this kind of exploration possible. When we turn to other areas of the country we have a terrible time matching Indian traditions with geological features because an outstanding natural formation is almost always isolated and bears no relationship to anything else.

A good example is the tradition, shared by many tribes, of the Devils Tower or, as the Sioux call it, the Bear's Lodge, in northeastern Wyoming. In this story three little girls were out picking flowers when they were pursued by three bears. They ran to the top of the nearest rock, which at that time was only a few feet above the ground. When the bears followed the girls up the rock, they appealed to the spirits for help, "... and the spirits caused the rock to grow. Higher and higher it rose."[24] When the bears left, the girls made a rope of their flowers and let themselves down.

Now everyone in their right mind knows that this story is devised to explain the peculiarities of the Bear's Lodge geological formation. Nevertheless, there is a curiosity here. If you have ever been to the Devils Tower, you will see fluted columns of volcanic material composing the giant structure and will be told by the Park Service guide that you are seeing a volcanic plug from which the surrounding strata have been eroded.

The curious thing about Devils Tower is that the rock appears to be badly eroded about a quarter of the way down and the remainder of the structure seems to have rock of a much cleaner and fresher look. It does look like the greater percentage of the rock has recently been pushed upward and that the original top part had been subjected to erosion for a considerable period of time. There are numerous volcanic cones southeast of Devils Tower, so some correlation could be made between the two sites. But it has always seemed to me that a small part of the tribal traditions, that the rock suddenly was raised straight out of the ground to new heights, may reflect some historical observation by Indians.

Probably more specific, and certainly more in tune with possible geological processes, is the tradition concerning the Badlands of South Dakota, one of the few earthquake traditions that we have among the Plains Indian nations. This area has in recent years received a decent amount of water, but in my youth it was a very stark landscape featuring a white clay formation that blinded you in the summertime. Originally, according to the Sioux, this area was a high fertile plain with fine grasses and trees. The Sioux enjoyed it as a hunting area, but one day they were invaded by a "fierce tribe from the mountains toward the setting sun. ..."[25] I suspect these invaders were Salish because their legends say they were more than a match for the three tribes of Dakotas and that they roamed far to the east in their hunting.

After suffering several severe defeats, the Sioux appealed to the spirits to help them. Their prayers were more than answered:

> The midday sky became as black as a midnight sky. Lightning flashed, seeming to come from the ground, and thunder rolled. Strange fires lighted the entire country with their flames. The earth shook. Where the western tribe had their camp, waves of land like waves of a great water rolled back and forth.[26]

Finally, the fires burned themselves out, leaving the devastation that we today know as the Badlands. Some versions of this story include modern camp descriptions such as horses and tipis, but the basic story of the origin of the Badlands seems historical. The geology of the area makes it difficult to judge the age of this event. George W. Kingsbury, in his monumental work *History of Dakota Territory*, published in 1915, noted that:

> It has been asserted by distinguished geologists that the basin of the Bad Lands is the ancient bed of a great coal field, the upper seam of which has been burned out by self-ignited fires, and the same layer underlies all the territory between the Missouri and the Yellowstone. In the early part of the last century the trappers and Indians told of the region of the Bad Lands being on fire, emitting an offensive smoke and the sound of rumbling thunder from the earth. These phenomena were mentioned by Lewis and Clark in 1805, and Hunt and McKenzie in 1811.[27]

Kingsbury also cited Charles Bates, who had surveyed the public lands in Dakota beginning in 1870. He in turn mentioned General Sully's remark in 1864 when chasing the Sioux and encountering the North Dakota Badlands: "Hell with the fires out and still smoking." Bates said the whole area was an extensive bed of lignite composed of vegetable materials extending from the Yellowstone River in eastern Montana clear across most of South Dakota and the lower half of North Dakota. "The smoke was caused from burning lignite beds, and some are still burning, or were when I was there twelve years ago," Bates noted.[28]

The area in question is in the neighborhood of six hundred square miles, which is an incredible tract of land to be covered by a layer of coal that sometimes gets as thick as thirty-five feet. An unnamed geologist cited by Kingsbury stated that there were "... trunks of fallen trees with the stumps of the same, sixteen feet in diameter, standing where they grew, having been found there by General Harney, Lieutenant Warren, General Sully and Professor Hayden and others."[29]

While attributing the perpetual fires to ad hoc lightning strikes is an acceptable explanation for the smoke, which can sometimes be seen today where the White River empties into the Missouri River, nevertheless the question arises why the land does not have as many smoking locations today as it ever did since virtually none of the areas has been exploited.

Two more recent Sioux traditions refer to this phenomena. On December 11, 1944, John Neihardt had one of his later interviews with Black Elk. The old medicine man told Neihardt that the White River was called Maka-izta, "Smoking Earth," because there was a place there where smoke came out.[30] The Sioux used to take their children to the place and tell them stories about when they lived in a land of volcanoes. The band carrying this tradition apparently did not know the story of the catastrophe in the Badlands. A Sioux elder, according to Ella E. Clark, recounted that Harney Peak, the highest mountain in the Black Hills, once issued columns of white smoke. "When my father was a young man," he recalled, "the peak was on fire."[31]

The Badlands represent a sizeable geographical area, which in its variety of fossil remains suggests a major catastrophe. If we simply enlarge the area and include the large animal burial grounds at the Agate Quarry on the Niobrara River in western Nebraska, we begin to see the scope of disaster that was once visited upon the northern Great Plains. A veritable cemetery of animals exists.

In fact the remains are so compacted and so spectacular that they are often cited in popular books, but the catastophic nature of these burials is rarely made clear. Richard Lull, in his book *Fossils,* wrote that "... the fossils are in such remarkable profusion in places as to form a veritable pavement of interlacing bones, very few of which are in their natural articulation with one another."[32] And he quoted some absolutely incredible figures which illustrate the magnitude of the deposits. Citing a block (actually a slab) of rock with a surface measuring 5.5 by 8 feet which contained 4,356 bones, he wrote that the estimate of the number of animals in the whole formation was "Diceratherium—16,400 skeletons, Moropus—500, and Dinohyus at 100 animals."[33] Surely the northern Great Plains is an area ripe for systematic re-evaluation. Hopefully the Sioux traditions about the region will be seen as contributions that point the way for new research topics.

Although some Christian fundamentalists believe that floods testify to the accuracy of the Old Testament, it seems obvious that the close identification of Indian traditions with specific land-marks places them in a distinctly different category. Only the few stories involving intense periods of rain *might* have some corre-lation and the Klamath period of prolonged rain must certainly deal with the aftermath of the volcanic eruption. One major benefit of validating these traditions would be to establish the ap-proximate time when certain tribes would have been occupying specific land areas. Linking the traditions of several tribes, such as we did in the Spokane flood story, enables us to see specific, if not spectacular, early events in North American geological history. Both scholars and disciplines have avoided cross-disciplinary cor-relation because of doctrinal considerations. Minimally, the rec-ognition of the historical basis for some of these accounts can help amend the artificial boundaries of occupancy areas which the Indian Claims Commission and other federal agencies have forced on many tribes.

Admitting the veracity of Indian accounts would also enable us to sketch out a fruitful area for research in establishing a more specific history of North America that could take the place of the present speculative chronology of "primitive man" with its endless phases of hypothetical occupations based on a few different pottery styles and some chipping rocks. Since there is some basis now for reducing the number of possible phases of the Ice Age, the resolution of this question may help us escape the silliness of the rise and fall of land bridges and ice corridors and make sense of human occupation of this continent.

10

At the Beginning

AS I HAVE ALREADY ARGUED, Western science today is akin to a world history which discusses only the Mediterranean peoples. Indeed, the institutionalization of knowledge in the academic setting has made status more important than accomplishments or ideas when determining the canon of truth that will give the best explanation of our planet. We are living in a strange kind of dark ages where we have immense capability to bring together information but when we gather this data, we pigeonhole it in the old familiar framework of interpretation, sometimes even torturing the data to make it fit.

Discordant facts and experiments are simply thrown away when they do not fit the prevailing paradigm. Once a theory such as the progression of human evolution, the Bering Strait land bridge, or the big-game hunters is published, it is treated as if it was *proven* and it is then popularized by people who rarely read the original documents, and vigorously defended by scholarly disciplines more fiercely than they would defend our country if called upon to do so. Alfred North Whitehead wrote in *Adventures of Ideas:*

> When the routine is perfect, understanding can be eliminated, except such minor flashes of intelligence as are required to deal with familiar accidents, such as a flooded mine, a prolonged drought, or an epidemic of influenza. A system will be the product of intelligence. But when the adequate routine is established, intelligence vanishes, and the system is maintained by a coordination of conditioned reflexes.[1]

No better description can be offered of the big-game hunter hypothesis when we see scholars seriously arguing that the Paleo-Indians killed the megafauna so quickly they didn't even leave large piles of bones.

On the whole, Indian tribes, faithful to their traditions, understand the origin of life on our planet as a creation. The stories may appear childish, but they do describe processes which may well parallel or describe segments of scenarios sometimes put forward by scientists. The perfect-symmetry proposals put forward by Heinz Pagels are not radically different from the Sioux creation story of Inyan the Rock. Some of the other tribal accounts have much in common with the binary theory of quanta-volution developed by Alfred de Grazia. The problem with the Indian traditions is that hardly any open-minded scientist has heard them, and an even fewer number know how to listen to Indian elders, catch the nuances of meaning, and be prepared to elicit the proper information from the story.

I am not a scientist and can only determine the state of our scientific knowledge by reading scholarly articles and popular writers to see what they say science knows. Obviously, from the objections I have raised in this book, a great deal of the current popular scientific beliefs and doctrines do not hold up to even the simplest critical review. I hope that the next generation of scholars, Indian and non-Indian, will force open any breaches I may have identified in the wall of scientific orthodoxy and make honest people out of scientists who are now afraid to publish their true beliefs and thoughts out of concern for peer conformity. To that end, therefore, I will now discuss areas where I believe good research and much difficult thinking will produce substantial breakthroughs in the years ahead.

CREATION

Apart from tribes having migration stories as descriptive of their origins, the majority of stories of origin suggest a creation in which people are given, simultaneous with their creation, an awareness that they have been created. These traditions often suggest that there was no essential spiritual/intellectual difference between people and animals. Some tribes report that an entity could change shape and experience what various birds and animals experience in that particular kind of body. Thus, stories relate that people and animals married each other. Peter Noyes, an elder on the Colville Reservation in northeastern Washington State, told Ella Clark: "Long ago—I don't know how long ago, the animals were the *people* of this country. They talked to one another the same as we do. And they married, too. That went on for many, many years, and then the world changed."[2] From the Sioux stories of compatible spirits I feel that this "marriage" must have been a blending of two kinds of individual spirits.

Human beings seem to be the focal point of communication in our world. Many traditions say that we cannot do anything very well except communicate, and consequently we are chosen to be the carriers of ceremonial thanksgiving activities on behalf of all other forms of life. We are not the only primate-shaped creatures, however, since there are peoples larger and smaller than we are, and some of these other peoples at one time coexisted with us, much as Cro-Magnon and Neanderthal are now known to have been contemporaries.

We seem to find cohesive biotic layers in our geological strata that represent whole life systems with virtually no evolutionary patterns present. It's as if a whole organic interlocking network including prey and predator and creatures dependent on a symbiotic relationship sprang into existence at once. If anything "punctuated," it experienced this spasm along with the creatures it devoured and the predators it fed. Then came the massive deposition of materials, quickly burying the system and preserving the fossils. And then a whole new biota came into existence. We always tread on the verge of "creation," but we cannot bring ourselves to face the probable "fact" of it.

Geologic time scales must inevitably be shortened by scientists. If we have previously allocated millions of years for the decline of the dinosaurs, and consensus is moving toward a quick destruction of these creatures by a meteor hit near Yucatán, do we really need those millions of years any longer? Time estimates on the age of Niagara Falls were once near one million years and are now figured in the thousands of years. Thus we may find ourselves reducing the time span for biotic life by substantial numbers, making it clear that things could not have evolved. I do not think that we will ever reach the short time favored by religious fundamentalists, but we may discover that the planet has a much shorter and more spectacular history than we thought possible.

THE EARLY CLIMATE

Many cultures look back on a golden age; many more see the history of the planet in terms of "ages" that are brought to an end by great catastrophes. Some thinkers have suggested that mountain chains rise quickly and dramatically during these events, building a chain of mountains in months and years rather than in millennia and millions of years. Since we know that the northern lands once had warmth of sufficient duration to grow corals in Greenland, and presumably daisies in Siberia, we have three choices in interpreting these facts: (1) continents shifted dramatically, moving some areas to warm climates and others to cold ones; (2) the planet once had a cloud canopy surrounding it, which provided uniform warm temperatures everywhere; or (3) prior to the tilt of the Earth's axis, there was uniform sunlight hitting the poles and equator, providing a uniform temperature.

It seems to me that a decent argument can be made for the idea that at one time, when people our size inhabited the Earth, the planet was shrouded with some kind of water canopy. While people could determine light and darkness, the canopy was too thick to produce clear images of the sun and moon. An additional feature of this time was that rain, snow, and thunderstorms, at least as we know them today, were not meteorological phenomena. Instead, the Earth was covered by a mist in which water continuously evaporated from the ground and precipitated again as mist.

Clarence Pickernell, a man with Quinault, Chehalis, and Cowlitz ancestors, said that "... when the world was young, the land east of where the Cascade Mountains now stand became very dry. This was in the early days before rains came to the earth. In the beginning of the world, moisture came up through the ground, but for some reason it stopped coming."[3] Genesis also records this phenomenon: "For the Lord God had not caused it to rain upon the earth, and there was not man to till the ground [Genesis 2:5]. But there went up a mist from the earth and watered the whole face of the ground [Genesis 2:6]."

We cannot tell which of the possible radical changes in the planet's composition these images might describe. This condition existed prior to the origin of the four rivers in Mesopotamia in the Old Testament, however. I have already noted that Sioux spoke of a time when there was no thunder and lightning and the rivers were dug by giants. So we are talking about a climatic condition in which there were no large rivers because there were insufficient amounts of rain to carve out river beds and provide flow. It is also proper to note in this connection that the appearance of the rainbow made a strong impression on ancient people, and that creates a dilemma because only the Hebrews connect the rainbow and Noah's flood. The subject of radically different climates is, however, a topic that will require some significant thought.

VOLCANISM

We can surmise that we are talking about some profound volcanism when some traditions speak of the world being destroyed by fire. Ager and Hodge both stated that the lava flows they encountered appeared as if they were fresh. The Sioux traveled to the White River to show their children the plumes of smoke and tell them about the time when they lived in a land of volcanoes. Our experience, and the Indian traditions we have used in this book, would suggest that volcanism was a local event, and our experiences in recorded history would seem to confirm that idea. Nevertheless it would be interesting to correlate the various studies on igneous rocks and see if there was a time when widespread volcanism was a major event on the planet, since the destruction of a locale would not necessarily be enough to cause people to believe that the world had indeed ended.

GEOLOGICAL STRATIGRAPHIC COLUMNS

Derek Ager raised many questions about the viability of the orthodox interpretations of deposition of strata, his major argument focusing on the fact that many strata regarded as sedimentary had a geographical expanse far in excess of what is possible. Thus the sandstones, limestones, and chalk that spanned whole continents and were found in some places around the world could not have been laid down gradually and calmly by incremental additions of sediment. He pointed out that contemporary measurements of deposition, including the ocean floors, were nowhere near the thickness required to produce the sedimentary rocks we find on the continents. According to Alfred de Grazia:

> If every different statum that was ever labeled were heaped up in its maximum deposited thickness, the pile would tower into the stratosphere. According to the accounts rendered of the world geologic column, there should be 400,000 feet, or 80 miles thick, of sediment. Furthermore, the heap should cover the globe, unless somebody has been digging rock from the oceans and carrying it up the continental shelves. For the ocean bottoms are scarcely sedimented.[4]

Creation should be excluded from explaining the sedimentary rock in that it posits a creator who periodically exterminates his biota, leaving extraterrestrial "dumps" of material from as-yet unidentified sources.

Reducing the millions of years required to lay down a bed of sandstone, shale, or limestone, depositing chalk and loess in a quick fly-by or fly-through of material, and coordinating the lava flows on a global basis could bring the strata containing flora and fauna much closer together. We could then explain "living fossils" and the presence of Jurassic and Triassic flora in the Pacific basin without crossing our fingers behind our backs. With some overlap of the periods having plentiful biospheres on the planet, the monsters of folklore and the giantism in strata would be seen as part of the early experiences of our species.

While geologists are very conservative, the major problem in this field is that it is so complicated. Hundreds of thousands of technical reports exist describing very complicated formations. But geologists are trapped in an unrealistic orthodox time scale by the paleontologists and evolutionists. Geology should properly be linked with astronomy and not with biology. The Shoemaker and Hale-Bopp Comets suggest that planets, especially ours, are not immune to cosmic catastrophes of unimaginable size and power. With the general acceptance of the Yucatán comet/meteor being the cause of the demise of the dinosaurs and the possibility that the Carolina bays and the Alaska lakes are remnants of a close fly-by of a comet/meteor, we can expect geology and astronomy to link ever closer and lead the way out of our present impasse.

LIVING FOSSILS

In December 1994, a marvelous discovery was made in Australia. David Noble of the Parks and Wildlife Service of that country discovered an isolated grove of a rainforest, a grove of what are enthusiastically described as "Wollemi pines," a pine tree with dense waxy foliage and knobby bark thought to have disappeared in the Jurassic period. Newspaper accounts suggested that this find was as significant as the discovery of the dawn redwood tree in China in 1944 and the coelacanth fossil fish off Madagascar in 1938.[5] Another instance of living fossils was the discovery of a Temnospondyl, a class of reptiles, found in 1978 by Anne Warren of Latrobe University.[6]

Everyone is pleased at the discovery of the Wollemi pines, but is it such a geological rarity? L. Taylor Hansen, in *The Ancient Atlantic*, wrote that "there is a botanical fact which demands explanation. The South Sea islands are covered by Triassic plants. These living fossils, not found elsewhere, demand some type of explanation."[7] Furthermore, she pointed out that by the end of the Jurassic, Japan and the Philippines had been isolated from the mainland and consequently have Jurassic flora.[8] So what explanation is given by scientists for these plants which seem to be immune from evolution? How can plants from 150 million years ago not have evolved into something else? How can they be sitting there as nice as can be when completely out of the geological, paleontological, and evolutionary time scale?

We have become so accustomed to accepting scientific authority that we have lost the ability to understand what they are saying. Scientific authorites, on the other hand, casually let fly remarks that drastically undercut their beloved evolutionary doctrines without even realizing what they have done. We have literally hundreds of "fossil creatures" from the past living with us today. In fact, when we begin to compile a list of the animals casually mentioned by scientists, it is alarming. Where is evolution?

Jacob Bronowski, writing about Africa in preparation for telling us about the wonderful fossil ancestors which Louis Leakey, Raymond Dart, and others have discovered there, discusses the alleged human remains, pointing out the great changes that took place over two million years. Then he says:

> ... naturally, we expect the animals of the Savannah also to have changed greatly. But the fossil record in Africa shows that this is not so. Look as the hunter does at the Topi antelope now. The ancestor of man that hunted its ancestor two million years ago would at once recognise the Topi today.[9]

If we had never seen a Topi prior to last Christmas, and had been assured by paleontologists that they have been dead for two million years, would we not be surprised and excited if one were discovered in unexplored Africa this past year? Does that make it a living fossil?

Robert Bakker, discussing a large tortoise, Colossochelys, which had been found "... everywhere in the Old World tropics, from

Kenya to Cape Provine to Java …," said that wherever it was found "… there was an accompanying rich fauna of big, modern mammals,"[10] indicating that it was probably not really a fossil, simply a species that recently became extinct. "Crocodiles," he noted, "first enter the chronicle of the rocks long after lizards but a few million years before dinosaurs."[11] And not just the crocs: "Alligators are dinosaur uncles—relatives of the direct ancestors of early dinosaurs—and as such they should be living representatives of the ancestral dinosaurs' forelimb arrangement."[12] In other words, many living creatures today should actually be classified as living fossils, representatives of remote geological eras, because they apparently arose in remote times and have somehow survived all subsequent geological changes to persist today.

Gordon Rattray Taylor, in *The Great Evolution Mystery*, notes the following: (1) Bees—"Bees preserved in amber from the Tertiary period are almost identical with living bees."[13] (2) Bacteria—"Since they reproduce themselves, in favorable conditions every twenty minutes, they might be expected to evolve faster than any other organism—but fossil bacteria going back to three and a half billion years, to the threshold of life itself, have been recovered and are virtually identical with modern forms."[14] (3) Salamanders— "… at the start of the Jurassic, there was a sudden and complete change to the modern type of Anuran, a change so successful that they have remained successful for 200 million years and have spread far and wide over the earth."[15] (4) Penguins—"No Antarctic fossil penguins have been found, but there are plenty in Australia, New Zealand, and Patagonia, going back to the Eocene; unfortunately they are very like modern penguins and tell us nothing about their origins."[16] (5) Oysters—"Many bivalves, such as oysters and mussels, have evolved very slowly, changing little in 400 million years." (6) And others—"the king crabs unchanged since the Triassic and the opossums."[17] And we must not forget the platypus and shark. A large number of familiar mammals and fish, as a matter of fact, have orthodox dates from around the Paleocene—12 million years ago.

It should not take a genius to recognize that the so-called antiquity of these creatures is illusory. We see hundreds of species in our modern world who are, in fact, survivors of previous Earth epochs. If we could find an honest scientist and have him or her make up a complete list of animals, fish, birds, reptiles, bacteria, and plants that "stopped evolving" millions of years ago and are found alive and kicking in the modern world, we would have a pretty good inventory of contemporary fauna and flora. The conifer genus *Metasequoia* was discovered in 1946 in China.[18] The deep-sea mollusk *Neopilin galatheae* was found in 1952 off the Mexican coast.[19] The list must be endless. It must be possible, and probably necessary to understand our situation, to collapse these millions of fictional years as much as we can and understand that our planet has a much different history than we have been told.

DINOSAURS

The discussion of living fossils was necessary because a number of tribal traditions describe creatures that may have been dinosaurs. In the worldview of orthodox science, such a suggestion is preposterous at first blush, but as we have seen, a number of fauna originated in very early times and the crocodile and alligator are said to have come on the scene before the dinosaurs flourished. The Tohono Oodham, formerly called the Papagos, live on a large reservation in southern Arizona adjoining the Mexican border. Most prominent on the eastern side of their lands is a large peak, Baboquivari, and this site plays an important role in their earliest traditions. An extremely large animal, personified as the spirit Etoi, was said to have inhabited the mountain and dominated the vicinity.

The Tohono Oodham are very secretive about this creature, but good authority suggests that it was some kind of dinosaur. Several years ago, Tohono Oodham from south of the border in northern Mexico and the American Tohono Oodham gathered for a ceremony to revive ancient customs. It is believed that one object of particular sacredness used in this ceremony was an unfossilized dinosaur bone from one of Etoi's personifications.

Again the Pacific Northwest peoples have a number of stories concerning oversized animals in their lakes and rivers. Since the current trend in dinosaur research suggests that these creatures, for the most part, were warm-blooded and had social and instinctual characteristics reminiscent of mammals of today, there is no reason to hesitate suggesting that some of these creatures, described as animals or large fish by observers, were surviving individuals of some presently classified dinosaur species. That is to say, humans and some creatures we have classified as dinosaurs were contemporaries.

The best-known story concerns the monster, known as Ogopogo, who lives in Lake Chelan on the east slope of the Cascades. Lake Chelan is fifty-five miles long, filling a glacial valley, and reaches depths of around 1,600 feet. Originally, the Washington area, according to a grandson of Chief Wapato of the Colville tribe, was a flat fertile grasslands prairie inhabited by grazing animals. A monster showed up and began devouring the animals, causing the Indians to go hungry. Twice they appealed to the Great Spirit, and he killed the monster, but it revived. The third time "... the Great Spirit struck the earth with his huge stone knife. All the world shook from his blow. A great cloud appeared over the plain." And when the cloud finally dissipated, the people could see that the land had been radically changed:

> Huge mountains rose on all sides of them. Among the mountains were canyons. Extending from the northwest to the southeast for two days' journey was a very deep canyon between high mountains.[20]

The monster was buried in this canyon, which was then filled with water to form Lake Chelan, and this lake was subject to sudden and intense wave disturbances, leading the people to say that the monster's tail was still alive and causing problems for them.

Why would we attempt to identify this creature as a dinosaur or comparable animal? Indians generally speak with a precise and literal imagery. As a rule, when trying to identify creatures of the old stories, they say they are "like" familiar neighborhood animals, but then carefully differentiate the perceived differences. I have found that if the animal being described was in any way comparable to modern animals, that similarity would be pointed out; the word "monster" would not be used.

Only in instances where the creature bears no resemblance to anything we know today will it be described as a monster. Since no dinosaur shape resembles any modern animal, and since the reports are to be given literal credibility, I must suggest that we are identifying a dinosaur. Thus, in the story of large animals at Pomme de Terre prairie in southwestern Missouri, a variant of the story suggests that the western animals were megafauna and the creatures who crossed the Mississippi and Missouri Rivers and invaded the lands of the megafauna were dinosaurs. The dinosaurs thus easily displace the familiar, perhaps Pleistocene, megafauna and move west, where we find their remains in the Rocky Mountains today.

In numerous places in the Great Lakes are found pictographs of a creature who has been described in the English translation as the "water panther." This animal has a saw-toothed back and a benign, catlike face in many of the carvings. Various deeds are attributed to this panther, and it seems likely that the pictographs of this creature, which are frequently carved near streams and lakes, are a warning to others that a water panther inhabits that body of water. The Sioux have a tale about such a monster in the Missouri River. According to reports, the monster had "... red hair all over its body ... and its body was shaped like that of a buffalo. It had one eye and in the middle of its forehead was one horn. Its backbone was just like a crosscut saw; it was flat and notched like a saw or cogwheel."[21] I suspect that the dinosaur in question here must be a stegosaurus.

Part of the original story implies that the creature was luminous at night, suggesting that someone was combining the memory of this creature with their first view of a steamboat. Frankly, that is a difficult argument to refute, although the Sioux phrase describing the steamboat is entirely different from the phrase used to describe this monster. I have asked various scholars who are familiar with the literature on the Sioux and many think it was some kind of animal. But their reasoning is somewhat askew. They believe that the Sioux saw a complete skeleton of this animal in the Badlands and then transferred the memory of the shape to account for a creature seen or imagined in the Missouri River.

The problem with this interpretation is that the Badlands are nearly 125 miles from the Missouri River, and consequently it would take a powerful imagination to make this kind of transference. I mentioned this anomaly in a speech at my own reservation, Standing Rock, in North Dakota, adjoining the Missouri River. After my speech a couple of the traditional people approached me and said that the next time I came, if I had time, they would take me to see the spot where the people last saw this creature, implying that it was still possible to see the animal during the last century before the reservations were established. I give their knowledge credence, in part because the location of the spot where the animal was last seen was upstream from the Badlands by about 150 miles.

On both the Standing Rock and Cheyenne River Reservations numerous dinosaur bones are found. After spring floods along the Grand and Moreau Rivers, people tell me that bones are washed out of the riverbanks. One of the best *Tyrannosaurus rex* skeletons we have was discovered on the Cheyenne River Reservation a few years back, on the lands of a tribal member. When private excavators tried to take the skeleton to Rapid City, a lawsuit ensued, and the tribe claimed the dinosaur as its heritage. So this part of South Dakota is most definitely dinosaur country.

In September 1996, *The Denver Post* did a feature article on the retirement of Dr. Bruce Erickson, the paleontologist who was in charge of the Wannagan Creek Quarry in the Badlands of North Dakota and who was responsible for discovering fifteen new taxa of flora and fauna. He spent twenty-six years at this location and had devised a gigantic wall map with drawings of thousands of bones on it. "That's only about a third of what we found," he was quoted as saying, "... the bones that were on the surface. There were twice as many beneath them."²² His first visit to the location produced about a dozen skulls lying on the surface. According to the article some alligators and crocodiles found in the quarry were contemporaries of the dinosaurs. But why would bones some 160 million years old be found on or very near the surface?

The dinosaur scenario sounds like overreaching to make a point about the longevity of American Indians, but there is a real basis for making the suggestion. In October and November 1924, a scientific expedition led by Samuel Hubbard, curator of archaeology at the Oakland Museum, Charles W. Gilmore, curator of vertebrate paleontology at the United States National Museum, and funded by the oil magnate about to be discredited, E. L. Doheny, went to Havasupai Canyon in northern Arizona to search for evidence of prehistoric man. Hubbard and Doheny had visited this area before, Doheny as a young prospector and Hubbard as a scientist.

Following the "Tobocobe Trail" to where it intersects with Lee Canyon, the party soon discovered what they described as "wall pictures," figures scratched long, long ago depicting the local fauna. The most spectacular of these pictures was one of a dinosaur, identified by them as *Diplodocus*, standing upright. The dinosaur figure was: "total height 11.2 inches, greatest width 7 inches, length of leg 3.8 inches, length of body 3.9 inches, width of body 3 inches, length of neck to top of curve 3.5 inches, length of tail (approximately) 9.1 inches, length of neck (approximately) 5.1 inches."²³ The desert "varnish," a covering caused by extreme age, had filled in the lines of the figure, indicating a significant age.

Just as spectacular, however, were other discoveries in the canyon. In Hubbard's words: "On the same wall with the dinosaur pictograph, and about 16 feet from it, we found a pictograph representing an animal which was evidently intended for an elephant, attacking a large Man. The elephant is striking the man on the top of his head with its trunk. The wavy line represents water into which the man has retreated up to his knees. Both arms are upraised and the fingers are visible on one hand. ... Because there are no tusks indicated our surmise is that it is a cow elephant."[24] This pictograph scene accurately depicts the manner in which scholars believe that man hunted the mammoth—an ambush at a waterhole; and in southern Arizona there are several sites which have man and mammoth remains together in an obvious hunting format, with butchering marks on the mammoth's bones.

The scientists of the expedition even tried to identify the species of mammoth represented in the picture. "Figuring the comparative scale of the two figures, if the elephant is identified as an *Elephas imperator* of California, it would be fourteen feet high and the man would also be fourteen feet high—a giant!"[25] In addition, in three different locations within the canyon, pictographs of ibex were found: "One group, showing a male and two females, was right under the elephant picture, but close to the ground."[26] Another picture showed men driving seven ibex and two deer into a canyon. Hubbard submitted the pictures to Roy Chapman Andrews and he identified the ibex from the knobs on the horns. And on part of the plateau above the canyon floor was found an ancient megalithic fortress.

This whole discussion would seem to beg the imagination, and certainly many scholars, when reminded of it, simply say, "Oh, that old thing," as if it had been satisfactorily explained decades ago. But it has not been explained in any satisfactory manner. Orthodox scholars simply omit it from consideration altogether, as they would with any bit of evidence that shakes the foundations of accepted scientific belief. Is it evidence that people and dinosaurs coexisted? I don't know, but I suspect so. Ibex fossils have never been found in North America, but there is no reason why lonely explorers from Africa could not have carved scenes from home in the canyon wall. It seems to me that the prevalence of dinosaur remains in South Dakota so near to the surface and this series of pictographs raise important questions which science has not bothered to ask.

RADIOCARBON DATING

Although very little evidence exists that Paleo-Indians were responsible for exterminating the mammoth, and no evidence exists regarding the other herbivores, the argument of Paul Martin and his supporters is that radiocarbon dating of human occupancy sites in the United States, particularly sites containing the famous Folsom points, concentrates at around 12,000 years before the present. Since these sites are scattered all over the western states, the implication is that the mass extermination was performed by the Indians. In *Pleistocene Extinctions,* Martin discusses various sites which have mammoth remains, discusses radiocarbon dating, and presents a chart which demonstrates that of fifty sites, half fall into the time frame in which he believes the great extinction occurred. Supporters of the theory often cite the radiocarbon dates as if a 50 percent score were proof "beyond a reasonable doubt."

What are the real facts about radiocarbon dating? We have seen that potassium-argon dating of volcanic rocks can go far afield, and we might begin to suspect that if a lava flow from 1800 in Hawaii can be K-Ar dated at 2.96 billion years, we are not getting much accuracy from these "scientific" measuring devices. Charles Ginenthal, in an excellent article on methods used for establishing the age of ancient artifacts entitled "Scientific Dating Methods in Ruins," points out that "J. Ogden, the director of a radiocarbon dating laboratory at Wesleyan University in Ohio, stated that the investigator is first asked what date he will accept for the material he brings to be dated; then, when a figure is obtained that comes near this date, it is duly reported—together with tolerance values— to make the test appear honest."[27] Ginenthal also cites several recent tests which should be more than enough to raise serious questions about the reliability of this way of dating materials:

1. In the *Antarctic Journal of the United States,* W. Dort wrote that freshly slaughtered seals, when subjected to radiocarbon analysis, are dated at 1,300 years old.

2. In *Science,* M. Keith and G. Anderson wrote that shells of living mollusks were dated at 2,300 years old.

3. In *The Physiology of Trees,* Bruno Haber wrote that wood from a growing tree was dated at 10,000 years old.[28]

This last test is important for our discussion because it seems that the tree was growing next to an airport which had a high level of carbon dioxide from airplane exhausts. (If a mammoth bone had been found nearby, would the tree be as old as the mammoth, or the mammoth as young as the tree?) Finally, Ginenthal quoted R. Stuckenrath from an article in the *Annals of the New York Academy of Science* regarding radiocarbon dating: "This whole blessed thing is nothing but thirteenth-century alchemy and it all depends upon which funny paper you read."[29]

Why don't scientists level with us? Why do they cite measuring techniques which have been found grossly inaccurate, as if they had absolute proof of their theories? Overkill fans piously mention that the debate is closed because radiocarbon dates have demonstrated a connection that cannot be denied. Yet they do not tell us that sometimes lab personnel have been told what dates scientists would prefer to have as the laboratory results. We can thank whatever deity inspires us that these people are in higher education and scientific labs and not in district attorneys' offices.

I discuss this subject because the next generation of people working on these problems, unless they are warned, will simply build on the errors of the famous personalities of the past. We will continue to wander in a darkness of our own making. But also, young Indians are being educated in increasing numbers, and when they try to discuss traditions of their people, even the most logical and credible stories can be turned aside by a mentor just saying, "But we have radiocarbon dates on this." We do indeed— we told the laboratory what dates we wanted and they gave them to us.

CLOVIS POINTS

A favorite game of anthropologists and archaeologists is to pretend that societies of early humans struggled for tens of thousands of years to make the simplest accommodations in the stone tools they made and used. Every change in the shape of arrowheads, pottery, or carving is understood as a radical departure from what had gone before. Thus marvelously poetic essays are written describing early, middle, and late Woodlands peoples, the Beaker and Red paint peoples, and so forth. If the same technique was applied to explain twentieth-century auto manufacturing we should have hundreds if not thousands of years intervening between the various models of Oldsmobiles. Today's small cars would be understood as a cultural revival of the business coupe of the late 1940s. All these divisions of sites into early, middle, and late stone crushers are, of course, wholly fictional devices to maintain the value of employing anthropologists when educational funds might be better spent.

According to scholars, the Paleo-Indians who marched out of Siberia across the Bering Strait and into America on a blood-lust campaign against megafauna had one superior technological innovation which spelled doom for the animals—the fluted point. No one knows the origin of this startling technology. There is but one fluted point in Siberia, so we must assume that they were developed after traversing the Bering Strait. Alex Krieger, in an essay in *Prehistoric Man in the New World*, said that the date for the invention of these points must be pushed back to around 15,000 years ago because of the immense distribution of these artifacts in the United States. He wrote:

> ... Mason [citation omitted] favors the central part of the eastern United States because Clovis points are most frequent there, and Witthof [citation omitted] believes that the oldest examples came from Pennsylvania.

And he argued that the old idea that the fluted point "reached North America from Asia via the Alaska steppingstone must be abandoned because it is not found in Asia."[30] Thus Martin and Diamond, eloquently chronicling the Clovis weaponry, will have to get their big-game hunters out of Tennessee, Pennsylvania, and Kentucky rather than bringing them from Asia.

Jared Diamond is quite firm in his identification of the Clovis people as the megafauna culprits. But the claim for Clovis is based, he says, on negative evidence: "... at excavated Clovis sites, conclusive evidence for artifacts made by other peoples has been found above but not below the level with Clovis tools; and *there are no irrefutable human remains with irrefutable pre-Clovis points anywhere in the New World south of the former Canadian ice sheet* [emphasis added]."[31] I have highlighted this statement because we need to see what Diamond thinks constitutes "irrefutable evidence." And he says:

> Mind you, there are dozens of claims of sites with pre-Clovis human evidence, but all of them are marred by serious questions about whether the material used for radiocarbon dating was contaminated by older carbon, or whether the dated material was really associated with the human remains, or whether the tools supposedly made by hand were just naturally shaped rocks. ...[32]

Applying what we know of radiocarbon dating, it is possible that samples were taken to a laboratory and the lab scientists were instructed to return dates of around 12,000 years ago. Tests were run and unfortunately the results showed something around 30,000 to 40,000 years ago, far too early to indict the people for mega-cide.

So what is the real state of the Clovis point? Alex Krieger explains that "... fluted points are found in all states west of the Rocky Mountains too, but for the most part they are surface finds and have not yet been connected with any particular culture pattern."[33] So this irrefutable evidence of an ice corridor invasion is scattered all over the West, lying on the surface of the ground, and difficult if not impossible to attach to any of the cultures that might have inhabited the area. Actually, the situation is worse than that. Waldo

Wedel writes: "... it is perhaps worth noting that many of the cutting, scraping and chopping tools of eight to ten thousand years ago differ little from those found on Plains Indian sites of the last five hundred years, and this is true also of such bone tools as awls, needles, etc."[34] So how would Martin, Ardrey, Diamond, or Bakker know whether the Clovis point they had just uncovered was the product of a Paleo-Indian of 15,000 years ago stalking a mammoth or simply a throwaway by a Sioux or Crow Indian in 1887 as they moved to their new reservation? They very likely wouldn't.

ALASKA

In this book's first edition I omitted Alaska, even though some articles had been written about the veracity of the oral tradition in that state. I am not familiar with the landscape and do not know any elders with whom the subject might have been discussed. In November 1995, the American Indian Science and Engineering Society (AISES) sponsored a conference on "Origins and Migrations," which dealt with the traditional knowledge of the tribes. Jana Harcharek, an Inupiat, stunned the conference with the stories of prehistory preserved by her people, describing Alaska as a temperate climate suddenly overwhelmed by a great catastrophe. In mid-February 1997, the AK MOKAKIT, a Canadian organization devoted to Native science and culture, held a conference that included some southern Athabascans—the White Mountain and Jicarilla Apaches sent delegates, as did the Navajos and Mescalero Apaches. As reported by Richard Pierce, the AISES representative, as stories and languages—actually dialects—were exchanged it became apparent that these people were related but that none of them had any traditions that related to the Bering Straits.

Some scholars have focused on the truth of the oral tradition in Alaska, granting the Natives credibility that is not accorded anyone in the lower forty-eight states. Perhaps the land requires that respect be paid to the Natives' understanding. Julie Cruikshank, in her article "Legend and Landscape: Convergence of Oral and Scientific Traditions in the Yukon Territory," quoted a journalist in 1890:

The Indian names of the mountains, lakes and rivers are natural land-
marks for the traveler, whoever he may be; to destroy these by substi-
tuting words of a foreign tongue is to destroy the natural guides.[35]

For much of Native tradition, particularly that part attached to
landscape, some measure of credibility has been recognized, making geomythology much more palatable for scholars.

Cruikshank concentrated on events near the shore and on more
recent activities, such as glacial advances and volcanic eruptions,
but she rejected earlier accounts of the White River volcanic explosion over a thousand years ago. Over a decade later, D. Wayne
Moodie, A.J.W. Catchpole, and Kerry Abel reviewed the tradition
and concluded that the Native tradition might very well have been
an eyewitness account.[36] Rory B. Egan recently wrote an essay in
which he compared the Mediterranean memories of the Thera
eruption with the White River volcano and inclined favorably toward recognizing the veracity in both traditions.[37]

The Alaska Natives have already held an important conference
on star knowledge at which a representation of Navajos from Navajo Community College, using specially made cylinders, were able
to compare Navajo constellations and Alaskan star knowledge. The
movement toward revival of traditional knowledge and its comparison with secular Western scientific knowledge is now well under way in the north and promises to produce a methodology for
gathering the data needed to verify Native traditions.

CONCLUSION

We do not know the real history of our planet and we know very
little about the historical experiences of the various societies and
races which constitute our species. This information is lacking
because our scholars and scientists are wedded to an outmoded
framework of interpretation and spend their time arranging facts
and evidence to fit these old ideas. Our popular science writers
seem to do very little thinking before their well-written, poetic,
but largely fictional accounts appear in the bookstores. Few of our

scholars or scientists seem able to recognize more than one possible explanation for data or phenomena, and they apparently hold
in great disdain all traditions except the one in which they have
grown up and received rewards.

Nothing can stop these people from filling us with more nonsense except an alert reading public with minds that are not inhibited by the prestigious degrees—a reading public that continues
to ask questions and demand sensible answers. Do our scientific
writers actually hold us in contempt? Do they feel that we can and
should be fed pablum so that "science" can continue? Sometimes
they resemble nothing more than priests of a dying religion, making up explanations ad hoc to defend outmoded articles of faith.

Most American Indians, I believe, were here "at the beginning"
and have preserved the memory of traumatic continental and
planetary catastrophes, keeping the information sometimes in
tales deliberately constructed to preserve as well as entertain.
When you visit with an elder, you will often get Coyote, Iktomi,
and Napi tales that are told in a manner designed to test you as
much as inform you. But if the elder thinks you are serious, or that
you already know something of the hidden knowledge, he will bow
his head for a moment, breathe deeply, and then begin to tell you
what his people have taught for thousands of years, giving as
closely as possible the literal description of the event. This kind of
information is generally not available to scholars who are on a
summer research grant and pester people about what they know.
That a particular story did not surface decades ago in the work of
a Western scholar should not discredit its veracity.

I hope this book will initiate discussions between traditional
people and scholars, and that it will provide a basis for the elders
to deal with overeducated younger Indians who have uncritically
accepted scientific folklore as fact. Nothing is more annoying than
listening to an educated Indian parroting what he or she has been
told in a lecture and discovering that tribal traditions have simply been thrown out the window without careful examination.
Many non-Indian scholars are ready to accord respect to tribal

traditions, but we have to be ready to engage in a free-for-all with them, critiquing their scientific folklore and making them provide the evidence and basis of their belief. Since the original publication of this book, I have been challenged many times about my disbelief in the Bering Strait migration, but I have yet to find one opponent who can provide me with one article or book that makes a convincing and reasonable case for believing in this nonsense.

Since American Indians have been unjustly accused of exterminating the megafauna, a good test of the question of respect might be to offer an equally silly accusation, based on a slim bit of evidence to counter the overkill thesis. G. Frederick Wright, one of the giants in Pleistocene geology, remarked on something that few people realize. The flora of Europe virtually disappeared during the Pleistocene, the same time period when the megafauna suffered a great reduction in the Western Hemisphere and, incidentally, on the other continents as well. Completely missing in Europe after the Pleistocene are what Wright called Atlantic American types—magnolia, Liriodendron, asimina, nogundo, Aesculus, leguminous trees such as locusts, honey-locusts, Gymnocladus, and Cladrastis, Nyssa, Liquidambar, Ericaceae, Bumelia, Catalpa, sassafras, Osage orange, hickory, walnut, hemlock, spruce, arborvitae, Taxodium, and Torreya.[38]

This list is most impressive and indicates that European flora in recent time was a rather drab scene, so that landing in America with its bountiful flora made the New World appear as a paradise rivaling the Garden of Eden. This list is also the equivalent of the listing of megafauna which Paul Martin believes were exterminated by Paleo-Indians. They were all missing or became extinct. Wright noted that:

> ... the capital fact is, that many and perhaps almost all of these genera of trees were well represented in Europe throughout the later Tertiary times. It had not only the same generic types, but in some cases even the same species, or what must pass as such, in the lack of recognizable distinctions between fossil remains and living analogues.

Probably the European Miocene forest was about as rich and various as is ours of the present day, and very like it. The Glacial period came and passed and these types have not survived there, nor returned.[39]

Like most of the megafauna bones, there is no evidence that humans had any part in the demise of these flora. However, as Jared Diamond felt a loosening of the rules of logic and evidence should be suspended in the case of the megafauna, so we must suspend the rules here also. It therefore seems painfully obvious that Europeans were the culprits of this dreadful floracide. With their new-found technology, the single blade point, they migrated up from Africa and west from the Caucasus and engaged in a frenzy of destruction in Europe, dwarfing anything seen in human history until their descendants marched through Tennessee and Kentucky at the close of the eighteenth century, always cutting far more hardwood trees than their homes needed.

Shouldn't we expect the same respect for our thesis accorded to the overkill thesis?

Notes

INTRODUCTION

1. Jacob Bronowski, *The Ascent of Man* (Boston: Little, Brown, 1973).

1
BEHIND THE BUCKSKIN CURTAIN

1. Will Durant, *The Age of Faith* (New York: Simon & Schuster, 1950),p. 775.

2. See *Lyng v. Northwest Indian Cemetery Protective Association,* 485 U.S. 439 (1988). Also *Employment Division, Department of Human Resources of Oregon v. Smith,* 110 S. Ct. 1595 (1990).

3. *Bowen v. Roy,* 476 U.S. 693 (1986).

2
SCIENCE AND THE ORAL TRADITION

1. Lawrence K. Altman, *New York Times Book Review,* May 16,1995, p. 3C.

2. Arnold Toynbee, *A Study of History* (New York: Oxford University Press, 1947), p. 39.

3. John Neihardt, *Black Elk Speaks* (Lincoln: University of Nebraska Press, 1961), p. 4.

4. Luther Standing Bear, *Land of the Spotted Eagle* (Boston: Houghton Mifflin, 1933), pp. 73–74.

5. *Newsweek,* February 3, 1992, pp. 53–54.

3
EVOLUTIONARY PREJUDICE

1. Michael Cremo and Richard Thompson, *Forbidden Archeology* (San Diego: Bhaktivedanta Institute, 1993), p. 259.

2. Ibid., pp. 412–13.

3. William Fix, *The Bone Peddlers* (New York: Macmillan, 1984), p. 123.

4. Ibid., p. 19.

5. William S. Laughlin, "Human Migration and Permanent Occupation in the Bering Sea Area." In *The Bering Land Bridge*, David M. Hopkins, ed. (Stanford, Calif.: Stanford University Press, 1967), p. 417.

6. Cremo and Thompson, p. 367.

7. Werner Muller, *America: The New World or the Old?* (Frankfurt am Main: Verlag Peter Lang, 1989), pp. 213–14.

8. Faye Flam, "Evidence Hints 2 Species of Early Man Coexisted." *Houston Chronicle*, January 6, 1997, p. 16D.

9. Quoted in Cremo and Thompson, p. 198.

10. Claude Levi-Strauss, "Saudades Do Brasil." *The New York Review of Books*, vol. xliii, no. 20, December 21, 1995, p. 20.

11. Cremo and Thompson, pp. 354–55.

12. Quoted in Cremo and Thompson, pp. 339–53, esp. 347.

13. The whole exchange of letters can be obtained in E. F. Greenman, "The Upper Paleolithic and the New World." *Current Anthropology* 4, no. 1 (February 1963): 41–71.

14. Muller, pp. 222–23.

4
LOW BRIDGE—EVERYBODY CROSS

1. H. Muller-Beck, "Migrations of Hunters on the Land Bridge in the Upper Pleistocene." In *The Bering Land Bridge*, David M. Hopkins, ed. (Stanford, Calif.: Stanford University Press, 1967), pp. 380–81.

2. Ibid., p. 374.

3. William S. Laughlin, "Human Migration and Permanent Occupation in the Bering Sea Area." In *The Bering Land Bridge*, David M. Hopkins,ed. (Stanford, Calif.: Stanford University Press, 1967), p. 421.

4. Ibid., p. 445.

5. Kazimierz Kowalski, "The Pleistocene Extinction of Mammals in Europe." In *Pleistocene Extinctions*, Paul S. Martin and H. E. Wright, Jr., eds. (New Haven, Conn.: Yale University Press, 1967), pp. 355–56.

6. N. K. Vereshchagin, "Primitive Hunters and Pleistocene Extinction in the Soviet Union." In *Pleistocene Extinctions*, Paul S. Martin and H. E. Wright, Jr., eds. (New Haven, Conn.: Yale University Press, 1967), pp. 372–73.

7. Stephen Taber, "Perennially Frozen Ground in Alaska: Its Origins and History." *Geological Society of America Bulletin* 54 (1943): 1,433–1,548, esp. 1,539.

8. Jared Diamond, "The American Blitzkrieg: A Mammoth Undertaking." *Discover* (June 1987): 82.

9. Ibid., p. 83.

10. Thomas Tyon, "Spirits." In *James Walker, Lakota Belief and Ritual*, Raymond DeMallie, ed. (Lincoln: University of Nebraska Press, 1980), p. 120.

11. Ella E. Clark, *Indian Legends of the Northern Rockies* (Norman: University of Oklahoma Press, 1966), p. 93.

12. Julian Steward, "Some Western Shoshoni Myths" (Washington, D.C.: Smithsonian Institution, *Bureau of American Ethnology, Bulletin 136, Anthropological Papers no. 31*, 1943), p. 299.

13. Frank Waters, *The Book of the Hopi* (New York: Penguin Books, 1963), pp. 39–40.

14. Donald Worster, *Dust Bowl* (New York: Oxford University Press, 1979), p. 72.

15. Taber, p. 1,530.

16. L. Taylor Hansen, *The Ancient Atlantic* (Amherst, Wisc.: Amherst Press, 1969), p. 257.

17. Ibid.

18. Robert T. Bakker, *The Dinosaur Heresies* (New York: William Morrow, 1986), p. 249.

19. Ibid., pp. 440–41.

20. Ibid., p. 443.

21. Ibid.

22. Stephen Jay Gould, *Ever Since Darwin* (New York: W. W. Norton, 1977), p. 90.

23. George Gaylord Simpson, "Mammals and Land Bridges." *Journal of the Washington Academy of Sciences* 30, no. 4 (April 1940): 159.

24. Ibid.

25. Ibid., p. 149.

26. Claude Levi-Strauss, "Saudades Do Brasil." *The New York Review of Books,* vol. xliii, no. 20, December 21, 1995, p. 20.

5
MYTHICAL PLEISTOCENE HIT MEN

1. Alex D. Krieger, "Early Man in the New World." In *Prehistoric Man in the New World,* Jesse D. Jennings and Edward Norbeck, eds. (Chicago: University of Chicago Press, 1963), p. 40.

2. Ibid., p. 45.

3. Robert T. Bakker, *The Dinosaur Heresies* (New York: William Morrow, 1986), p. 78.

4. Robert Ardrey, *The Hunting Hypothesis* (New York: Atheneum, 1976), p. 109.

5. Ibid., pp. 9–10.

6. Kazimierz Kowalski, "The Pleistocene Extinction of Mammals in Europe." In *Pleistocene Extinctions,* Paul S. Martin and H. E. Wright, Jr., eds. (New Haven, Conn.: Yale University Press, 1967), p. 354.

7. Robert Ebisch, "Megafauna Mystery." *Seattle Times,* May 21, 1990, p. 81.

8. Paul S. Martin, "Prehistoric Overkill." In *Pleistocene Extinctions,*

Paul S. Martin and H. E. Wright, Jr., eds. (New Haven, Conn.: Yale University Press, 1967), p. 115.

9. Ibid., p. 102.

10. Ibid., p. 88.

11. Ibid.

12. Letter from James Wright to John Bartram, August 22, 1762, quoted by George Gaylord Simpson, "The Discovery of Fossil Vertebrates in North America." *Journal of Paleontology* 17 (1943): 36.

13. James Hester, "The Agency of Man in Animal Extinctions." In *Pleistocene Extinctions*, Paul S. Martin and H. E. Wright, Jr., eds. (New Haven, Conn.: Yale University Press, 1967), pp. 179–80.

14. Ibid., p. 181.

15. N. K. Vereshchagin, "Primitive Hunters and Pleistocene Extinction in the Soviet Union." In *Pleistocene Extinctions*, Paul S. Martin and H. E. Wright, Jr., eds. (New Haven, Conn.: Yale University Press, 1967), p. 373.

16. Ibid., p. 380.

17. Ibid., p. 395.

18. Ibid., p. 388.

19. Arthur J. Jelinek, "Man's Role in the Extinction of Pleistocene Faunas." In *Pleistocene Extinctions*, Paul S. Martin and H. E. Wright, Jr., eds. (New Haven: Yale University Press, 1967), p. 195.

20. Ibid., p. 198.

21. Paul S. Martin, "Who or What Destroyed Our Mammoths?" In *Megafauna and Man: Discovery of America's Heartland*, Larry Agenbroad, ed. (Flagstaff: Northern Arizona University, 1990), pp. 116–17.

6
THE CORPORA DELICTI AND OTHER MATTERS

1. Jared Diamond, "The American Blitzkrieg: A Mammoth Undertaking." *Discover* (June 1987): 84.

2. Ibid.

3. Ibid., p. 88.

4. Fred Warshofsky, *Doomsday: The Science of Catastrophism* (New York: Readers Digest Press, 1977), p. 211.

5. Rev. D. Gath Whitley, "The Ivory Islands in the Arctic Ocean." *Philosophical Society of Great Britain* 12 (1910): 40–41.

6. Ibid., p. 43.

7. Ibid., p. 42.

8. Ibid., p. 45.

9. N. K. Vereshchagin, "Primitive Hunters and Pleistocene Extinction in the Soviet Union." In *Pleistocene Extinctions,* Paul S. Martin and H. E. Wright, Jr., eds. (New Haven, Conn.: Yale University Press, 1967), p. 388.

10. Stephen Taber, "Perennially Frozen Ground in Alaska: Its Origins and History." *Geological Society of America Bulletin* 54 (1943): 1,488.

11. Frank C. Hibben, "Evidence of Early Man in Alaska." *American Antiquity* 8 (1943): 256.

12. Taber, p. 1,490.

13. Ibid.

14. Letter of June 29, 1993, from Paul Martin to Vine Deloria, Jr., reads: "Whatever caused the frozen mucks of Siberia to be so rich in bones and even hide, hair, tissue and gut contents, it was not a sudden quick freeze caused by some Velikovskian accident."

15. Geoffrey Woodard and Leslie Marcus, "Rancho La Brea Fossil Deposits: A Re-evaluation from Stratigraphic and Geological Evidence." *Journal of Paleontology* 47, no. 1 (January 1973): 56.

16. Ibid.

17. Ibid., p. 58.

18. John C. Merriam, "The Fauna of Rancho La Brea. Part I: Occurrence." *Memoirs of the University of Calfornia* [Berkeley] 1, no. 2 (1911): 213.

19. Chester Stock, *Rancho La Brea: A Record of Pleistocene Life in California,* 7th ed. (Los Angeles: Natural History Museum, Science Series No. 37, 1992), p. 24.

20. Ibid., p. 32.

21. Paul S. Martin, "Prehistoric Overkill." In *Pleistocene Extinctions*, Paul S. Martin and H. E. Wright, Jr., eds. (New Haven, Conn.: Yale University Press, 1967), p. 101.

22. Thomas Jefferson, *Notes on the State of Virginia* (Boston: Thomas & Andrews, J. West, West & Greenleaf et al., 1801), pp. 59–61. Quoted in Ludwell H. Johnson III, "Men and Elephants in America," *Scientfic Monthly* 75 (October 1952): 217.

23. Ibid.

24. W. D. Strong, "North American Indian Traditions Suggesting a Knowledge of the Mammoth." *American Anthropologist* 36 (1934): 86.

25. Ludwell H. Johnson III, "Men and Elephants in America," *Scientfic Monthly* 75 (October 1952): 217.

26. Ibid., p. 215.

27. F. T. Siebert, Jr., Discussion and Correspondence: "Mammoth or 'Stiff-Legged Bear,'" *American Anthropologist* 39 (1937): 724–25.

28. M. F. Ashley Montagu, "An Indian Tradition Relating to the Mastodon." *American Anthropologist* 46 (1944): 568–69.

29. Ibid., p. 570.

30. Jane C. Beck, "The Giant Beaver: A Prehistoric Memory?" *Ethnohistory* 19 (1972): 119.

31. Johnson, p. 219.

7
CREATURES THEIR OWN SIZE

1. Ella E. Clark, *Indian Legends of the Pacific Northwest* (Berkeley: University of California Press, 1952), p. 15.

2. Ibid., p. 16.

3. Ella E. Clark, *Indian Legends of the Northern Rockies* (Norman: University of Oklahoma Press, 1966), pp. 113–14.

4. David M. Pendergast and Clement W. Meighan, "Folk Traditions as Historical Fact: A Paiute Example." *Journal of American Folklore* 72 (1959): 128–33.

5. Lord Raglan, "Folk Traditions as Historical Fact," Notes and Queries, *Journal of American Folklore* 73 (1960): 59.

6. Jared Diamond, "The American Blitzkrieg: A Mammoth Under-taking." *Discover* (June 1987): 84.

7. Ibid., p. 88.

8. Fred Warshofsky, *Doomsday: The Science of Catastrophism* (New York: Readers Digest Press, 1977), pp. 212–13.

9. William Ellis Edwards, "The Late Pleistocene Extinction and Diminution in Size of Many Mammalian Species." In *Pleistocene Extinctions,* Paul S. Martin and H. E. Wright, Jr., eds. (New Haven, Conn.: Yale University Press, 1967), p. 151.

10. Ibid., p. 147.

11. J. F. H. Claiborne, *Mississippi: As a Province, Territory, and State* (Baton Rouge: Louisiana State University Press, 1964), p. 484.

12. Edwards, p. 149.

13. Ibid.

14. Ibid., p. 150.

15. Frances Densmore, *Teton Sioux Music* (Washington, D.C.: Bureau of American Ethnology, Smithsonian Institution, 1918), pp. 137–38.

16. Clark, *Indian Legends of the Pacific Northwest,* p. 64.

17. Charles Eastman, *Indian Boyhood* (New York: Dover, 1971), pp. 164–65.

18. Loren C. Eiseley, "Archaeological Observations on the Problem of Post-Glacial Extinctions." *American Antiquity* 8, no. 3 (January 1943): 217.

19. Chester Stock, *Rancho La Brea: A Record of Pleistocene Life in California,* 7th ed. (Los Angeles: Natural History Museum, Science Series No. 37, 1992), p. 46.

20. Donald Patten, "A Comprehensive Theory on Aging, Gigantism and Longevity." In *Catastrophism and Ancient History* 2, pt. 1 (August 1979): 29. Quoted from Flavius Josephus, *The Antiquities of the Jews,* trans. William Whiston (Bridgeport, Conn.: Sherman, 1828), pp. 87–88.

21. Patten, p. 49.

22. Derek Ager, *The Nature of the Stratigraphic Record,* 1st ed. (New York: John Wiley & Sons, 1973), p. 25.

23. Stock, p. 37.

24. E. A. Vangengeim, "Quaternary Mammalian Faunas of Siberia and North America." In *The Bering Land Bridge*, David M. Hopkins, ed. (Stanford, Calif.: Stanford University Press, 1967), p. 285.

25. Quoted in Herbert Asbury, *The French Quarter* (Garden City, N.Y.: Garden City Publishing, 1938), p. 30.

26. Patten, p. 55.

8
GEOMYTHOLOGY AND
THE INDIAN TRADITIONS

1. Derek Ager, *The New Catastrophism* (London: Cambridge University Press, 1993), p. 14.

2. Ibid., pp. 1–2.

3. Ibid., p. 7.

4. Ibid., p. 9.

5. Ibid., p. 57.

6. Ibid., p. 43.

7. Ibid., p. 137.

8. Ibid., p. 156.

9. Ibid., p. 159.

10. Ibid., p. 165.

11. Philip B. King, *The Evolution of North America* (Princeton, N.J.: Princeton University Press, 1977), p. 131.

12. Jacob Bronowski, *The Ascent of Man* (Boston: Little, Brown, 1973), p. 66.

13. Dorothy Vitaliano, *Legends of the Earth* (Bloomington: Indiana University Press, 1973), p. 1.

14. Ella E. Clark, *Indian Legends of the Pacific Northwest* (Berkeley: University of California Press, 1952), pp. 23–24.

15. Vitaliano, p. 52.

16. Ibid.

17. Clark, *Indian Legends of the Pacific Northwest*, pp. 15–16.

18. Vitaliano, p. 126.

19. Ibid., pp. 126–27.

20. Lyman J. Briggs, "When Mt. Mazama Lost Its Top." *National Geographic* 122 (July–December 1962): 128–33.

21. Clark, *Indian Legends of the Pacific Northwest*, p. 53.

22. Ibid.

23. Ibid., p. 54.

24. Ibid., pp. 53–55. I have used extensive paraphrasing here because the original tale seemed to have many important points but reproducing the whole story would have been tedious for the reader.

25. Edwin T. Hodge, *Mount Multnomah* (Eugene: University of Oregon, 1925), p. 1.

26. Ibid., p. 118.

27. Clark, *Indian Legends of the Pacific Northwest*, pp. 13–14.

28. Hodge, p. 45.

29. Michael Cremo and Richard Thompson, *Forbidden Archeology* (San Diego: Bhaktivedanta Institute, 1993), p. 694.

30. Harold S. Colton, "A Possible Hopi Tradition of the Eruption of Sunset Crater." *Museum Notes* (Flagstaff, Ariz.: Museum of Northern Arizona) 5, no. 4 (October 1932): 23.

31. Ekkehart Malotki with Michael Lomatuway'ma, *Earth Fire: A Hopi Legend of the Sunset Crater Eruption* (Flagstaff, Ariz.: Northland Press, 1987), pp. 79–81.

9
FLOODS, LAKES, AND EARTHQUAKES

1. Ella E. Clark, *Indian Legends of the Pacific Northwest* (Berkeley: University of California Press, 1952), p. 44.

2. Ibid., p. 45.

3. James G. Swan, *The Indians of Cape Flattery* (Washington, D.C.: Smithsonian Contributions to Knowledge, no. 220, 1869), p. 57.

4. Ella E. Clark, *Indian Legends of the Northern Rockies* (Norman: University of Oklahoma Press, 1966), p. 39.

5. Clark, *Indian Legends of the Pacific Northwest*, p. 39.

6. Ibid., pp. 161–62.

7. John Eliot Allen and Marjorie Burns, with Samuel Sargent, *Cataclysms on the Columbia* (Portland, Ore.: Timber Press, Scenic Trips to the Northwest's Geologic Past, no. 2, 1986), p. 104.

8. Ibid., pp. 86–87.

9. Ibid., p. 84.

10. Ibid., p. 83.

11. Ibid., pp. 83–84.

12. Derek Ager, *The New Catastrophism* (London: Cambridge University Press, 1993), p. 21.

13. Allen et al., p. 88.

14. Ibid., p. 109.

15. I owe a debt of thanks to Deward Walker for allowing me to use this version of the story from an unpublished article of his. I knew Alex Sherwood, and his knowledge and integrity were unquestionable.

16. Clark, *Indian Legends of the Pacific Northwest*, pp. 74–75.

17. Allen et al., p. 106.

18. C. Warren Hunt, "Catastrophic Termination of the Last Wisconsin Ice Advance: Observations in Alberta and Idaho." *Bulletin of Canadian Petroleum Geology* 25, no. 3 (1977): 468.

19. Clark, *Indian Legends of the Northern Rockies*, p. 40.

20. Ibid., p. 90.

21. Ibid.

22. Clark, *Indian Legends of the Pacific Northwest*, p. 172.

23. Clark, *Indian Legends of the Northern Rockies*, p. 200.

24. Ibid., p. 306.

25. Ibid., p. 309.

26. Ibid., pp. 309–10.

27. George W. Kingsbury, *History of Dakota Territory*, vol. 1 (Chicago: St. Clarke Publishing, 1915), p. 869.

28. Ibid., p. 870.

29. Ibid., p. 868.

30. Raymond J. DeMallie, *The Sixth Grandfather* (Lincoln: University of Nebraska Press, 1984), p. 366.

31. Clark, *Indian Legends of the Northern Rockies*, p. 307.

32. Richard Lull, *Fossils* (New York: The University Society, 1931), pp. 31–32.

33. Ibid., p. 34.

10

AT THE BEGINNING

1. Alfred North Whitehead, *Adventures of Ideas* (New York: Macmillan, 1933), p. 96.

2. Ella E. Clark, *Indian Legends of the Pacific Northwest* (Berkeley: University of California Press, 1952), p. 81.

3. Ibid., p. 25.

4. Alfred de Grazia, *Chaos and Creation* (Princeton, N.J.: Metron Publications, 1981), p. 35.

5. Peter Spielmann, "Botanists Giddy over Rare Find in Australia." *The Denver Post*, December 15, 1994, p. 11A.

6. Gordon Rattray Taylor, *The Great Evolution Mystery* (New York: Harper & Row, 1983), p. 89.

7. L. Taylor Hansen, *The Ancient Atlantic* (Amherst, Wisc.: Amherst Press, 1969), p. 201.

8. Ibid., p. 219.

9. Jacob Bronowski, *The Ascent of Man* (Boston: Little, Brown, 1973), p. 26.

10. Robert T. Bakker, *The Dinosaur Heresies* (New York: William Morrow, 1986), p. 77.

11. Ibid., p. 23; note also p. 160, where Bakker observes that water-loving turtles and crocodiles evolve most slowly, changing so little on average through geological time that a single genus can be followed for 30 million years or more.

12. Ibid., pp. 207–08.

13. Taylor, pp. 24–26.

14. Ibid., pp. 26–27.

15. Ibid., p. 62.

16. Ibid., p. 74.

17. Ibid., p. 84.

18. Editorial Miscellany, *American Scientist* 36 (October 1948): 490.

19. Bentley Glass, "New Missing Link Discovered," *Science* 126 (July 1957): 158–59.

20. Clark, *Indian Legends of the Pacific Northwest*, p. 71.

21. Ella E. Clark, *Indian Legends of the Northern Rockies* (Norman: University of Oklahoma Press, 1966), p. 301.

22. Jane E. Brody, "It's Been 26 Years for Fossil Hound at N.D. Site." *The Denver Post*, September 8, 1996, pp. 32A–33A.

23. Samuel Hubbard, *The Doheny Scientific Expedition to the Hava Supai Canyon, Northern Arizona*, p. 9. A copy of this privately printed field report exists in the Peabody Museum of American Archaeology and Ethnology, Harvard University, a gift from Dr. A. V. Kidder, a prominent scholar in his own right.

24. Ibid., p. 15.

25. Ibid.

26. Ibid., p. 17.

27. Charles Ginenthal, "Scientific Dating Methods in Ruins." *The Velikovskian* 2, no.1 (1994): 53.

28. Ibid.

29. Ibid., p. 54.

30. Alex D. Krieger, "Early Man in the New World." In *Prehistoric Man in the New World,* Jesse D. Jennings and Edward Norbeck, eds. (Chicago: University of Chicago Press, 1963), p. 55.

31. Jared Diamond, "The American Blitzkrieg: A Mammoth Undertaking." *Discover* (June 1987): 84.

32. Ibid.

33. Krieger, p. 35.

34. Waldo R. Wedel, "The Great Plains." In *Prehistoric Man in the New World,* Jesse D. Jennings and Edward Norbeck, eds. (Chicago: University of Chicago Press, 1963), p. 198.

35. Julie Cruikshank, "Legend and Landscape: Convergence of Oral and Scientific Traditions in the Yukon Territory." *Arctic Anthropology,* 18, no. 2 (1981): 79.

36. D. Wayne Moodie, A.J.W. Catchpole, and Kerry Abel, "Northern Athapascan Oral Traditions and the White River Volcano." *Ethnohistory* 39, no. 2 (spring 1992): 148–171.

37. Rory B. Egan, "Ex Occidente Lux: Catastrophic Volcanism in Greek and *Dene* Oral Tradition." In *nikotwâsik iskwâhtêm, pâskihtêpayih, Studies in Honor of H. C. Wolfart,* John D. Nichols and Arden C. Ogg, eds. Memoir 13, Algonquian and Iroquoian Linguistics, 1996, pp. 196–209.

38. G. Frederick Wright, *The Ice Age in North America* (New York: D. Appleton, 1890), p. 378.

39. Ibid.

Bibliography

Agenbroad, Larry, ed. *Megafauna and Man: Discovery of America's Heartland.* Flagstaff: Northern Arizona University, 1990.

Ager, Derek. *The Nature of the Stratigraphic Record,* 1st ed. New York: John Wiley & Sons, 1973.

———. *The New Catastrophism.* London: Cambridge University Press, 1993.

Allen, John Eliot, and Marjorie Burns, with Samuel Sargent. *Cataclysms on the Columbia.* Portland, Ore.: Timber Press, Scenic Trips to the Northwest's Geologic Past, no. 2, 1986.

Altman, Lawrence K. *The New York Times Book Review,* May 16, 1995, p. 3C.

Anderson, Elaine. "Who's Who in the Pleistocene: A Mammalian Bestiary." In *Quaternary Extinctions,* Paul S. Martin and Richard G. Klein, eds. Tucson: University of Arizona Press, 1984, pp. 40–89.

Andrews, Roy Chapman. *On the Trail of Ancient Man.* Garden City, N.Y.: Garden City Publishing, 1926.

Ardrey, Robert. *The Hunting Hypothesis.* New York: Atheneum, 1976.

Asbury, Herbert. *The French Quarter.* Garden City, N.Y.: Garden City Publishing, 1938.

Bakker, Robert T. *The Dinosaur Heresies.* New York: William Morrow, 1986.

Beck, Jane C. "The Giant Beaver: A Prehistoric Memory?" *Ethnohistory* 19 (1972): 109–22.

Begley, Sharon, with Elizabeth Ann Leonard. "Take Two Roots; Call Me ..." *Newsweek,* February 3, 1992, pp. 53–54.

Bretz, J. Harlan. "The Lake Missoula Floods and the Channeled Scabland." *Journal of Geology* 77 (1969): 505–03.

Briggs, Lyman J. "When Mt. Mazama Lost Its Top." *National Geographic* 122 (July–December 1962): 128–33.

Brody, Jane E. "It's Been 26 Years for Fossil Hound at N.D. Site." *The Denver Post,* September 8, 1996, p. 32A–33A.

Bronowski, Jacob. *The Ascent of Man.* Boston: Little, Brown, 1973.

Claiborne, J. F. H. *Mississippi: As a Province, Territory, and State.* Baton Rouge: Louisiana State University Press, 1964.

Clark, Ella E. *Indian Legends of the Pacific Northwest.* Berkeley: University of California Press, 1952.

————. *Indian Legends of the Northern Rockies.* Norman: University of Oklahoma Press, 1966.

Colton, Harold S. "A Possible Hopi Tradition of the Eruption of Sunset Crater." *Museum Notes* (Flagstaff, Ariz.: Museum of Northern Arizona) 5, no. 4 (October 1932): 23.

Cremo, Michael, and Richard Thompson. *Forbidden Archeology.* San Diego: Bhaktivedanta Institute, 1993.

Cruickshank, Julie. "Legend and Landscape: Convergence of Oral and Scientific Traditions in the Yukon Territory." *Arctic Anthropology* 18, no. 2 (1981): 79.

de Grazia, Alfred. *Chaos and Creation.* Princeton, N.J.: Metron Publications, 1981.

DeMallie, Raymond J., ed. *James Walker, Lakota Belief and Ritual.* Lincoln: University of Nebraska Press, 1980

DeMallie, Raymond J. *The Sixth Grandfather.* Lincoln: University of Nebraska Press, 1984.

Densmore, Frances. *Teton Sioux Music.* Washington, D.C.: Bureau of American Ethnology, Smithsonian Institution, 1918.

Diamond, Jared. "The American Blitzkrieg: A Mammoth Undertaking." *Discover* (June 1987): 82–88.

Durant, Will. *The Age of Faith.* New York: Simon & Schuster, 1950.

Eastman, Charles. *Indian Boyhood.* New York: Dover, 1971.

Ebisch, Robert. "Megafauna Mystery." *Seattle Times*, May 21, 1990, p. 81.

Editorial Miscellany, *American Scientist* 36 (October 1948): 492–94.

Edwards, William Ellis. "The Late Pleistocene Extinction and Diminution in Size of Many Mammalian Species." In *Pleistocene Extinctions*, Paul S. Martin and H. E. Wright, Jr., eds. New Haven, Conn.: Yale University Press, 1967, pp. 141–54.

Egan, Rory B. "Ex Occidente Lux: Catastrophic Volcanism in Greek and Dene Oral Tradition." *nikotwâsik iskwahtêm, pâskihtêpayih! Studies in Honor of H. C. Wolfart*, John D. Nichols and Arden C. Ogg, eds. Memoir 13, Algonquin and Iroquoian Linguistics, 1996.

Eiseley, Loren C. "Archaeological Observations on the Problem of Post-Glacial Extinctions." *American Antiquity* 8, no. 3 (January 1943): 209–17.

———. "The Fire-Drive and the Extinction of the Terminal Pleistocene Fauna." *American Anthropologist* 48 (1946): 54–59.

Erman, Adolph. *Travels in Siberia Including Excursions Northwards, Down to the Obi, to the Polar Circle, and Southwards to the Chinese Frontier.* Translated from the German by W. D. Cooley. Philadelphia: Lea and Blanchard, 1850.

Fix, William. *The Bone Peddlers.* New York: Macmillan, 1984.

Flam, Faye, "Evidence Hints 2 Species of Early Man Coexisted." *Houston Chronicle*, January 6, 1997, p. 16D.

Ginenthal, Charles. "Scientific Dating Methods in Ruins." *The Velikovskian* 2, no. 1 (1994): 50–79.

Glass, Bentley. "New Missing Link Discovered." *Science* 126 (July 1957): 158–59.

Goddard, Pliny Earle. "The Beaver Indians." Vol. 10, pt. 4, of *Anthropological Papers of the American Museum of Natural History.* New York: American Museum of Natural History, 1916.

Gould, Stephen Jay. *Ever Since Darwin.* New York: W.W. Norton, 1977.

Greenman, E. F. "The Upper Paleolithic and the New World." *Current Anthropology* 4, no. 1 (February 1963): 41–91.

Hansen, L. Taylor. *The Ancient Atlantic.* Amherst, Wisc.: Amherst Press, 1969.

Hester, James. "The Agency of Man in Animal Extinctions." In *Pleistocene Extinctions,* Paul S. Martin and H. E. Wright, Jr., eds. New Haven, Conn.: Yale University Press, 1967, pp. 169–92.

Hibben, Frank C. "Evidence of Early Man in Alaska." *American Antiquity* 8 (1943): 254–59.

Hodge, Edwin T. *Mount Multnomah.* Eugene: University of Oregon, 1925.

Hopkins, David M., ed. *The Bering Land Bridge.* Stanford, Calif.: Stanford University Press, 1967.

Hubbard, Samuel. *The Doheny Scientific Expedition to the Hava Supai Canyon, Northern Arizona.* Private printing, copy in the Peabody Museum of American Archaeology and Ethnology, Harvard University, Cambridge.

Hunt, C. Warren. "Catastrophic Termination of the Last Wisconsin Ice Advance: Observations in Alberta and Idaho." *Bulletin of Canadian Petroleum Geology* 25, no. 3 (1977): 468.

Jefferson, Thomas. *Notes on the State of Virginia.* Boston: Thomas & Andrews, J. West, West & Greenleaf, et al., 1801.

Jelinek, Arthur J. "Man's Role in the Extinction of Pleistocene Faunas." In *Pleistocene Extinctions,* Paul S. Martin and H. E. Wright, Jr., eds. New Haven, Conn.: Yale University Press, 1967, pp. 193–200.

Jennings, Jesse D., and Edward Norbeck, eds. *Prehistoric Man in the New World.* Chicago: University of Chicago Press, 1963.

Johnson, Ludwell H., III. "Men and Elephants in America." *Scientific Monthly* 75 (October 1952): 215–21.

Kelley, Klara Bonsack, and Harris Francis. *Navajo Sacred Places.* Bloomington: Indiana University Press, 1994.

King, Philip B. *The Evolution of North America.* Princeton: Princeton University Press, 1977.

Kingsbury, George W. *History of Dakota Territory,* 2 vols. Chicago: S. J. Clarke Publishing, 1915.

Kowalski, Kazimierz. "The Pleistocene Extinction of Mammals in Europe." In *Pleistocene Extinctions,* Paul S. Martin and H. E. Wright, Jr., eds. New Haven, Conn.: Yale University Press, 1967, pp. 346–64.

Krieger, Alex D. "Early Man in the New World." In *Prehistoric Man in the New World,* Jesse D. Jennings and Edward Norbeck, eds. Chicago: University of Chicago Press, 1963, pp. 23–81.

Lankford, George E. "Pleistocene Animals in Folk Memory." *Journal of American Folklore* 93 (July–September 1980): 293–304.

Laughlin, William S. "Human Migration and Permanent Occupation in the Bering Sea Area." In *The Bering Land Bridge,* David M. Hopkins, ed. Stanford, Calif.: Stanford University Press, 1967, pp. 409–50.

Levi-Strauss, Claude. "Saudades Do Brasil." *The New York Times Review of Books,* vol. xliii, no. 20, December 21, 1995, pp. 19–21.

Lull, Richard. *Fossils.* New York: The University Society, 1931.

Malotki, Ekkehart, with Michael Lomatuway'ma. *Earth Fire: A Hopi Legend of the Sunset Crater Eruption.* Flagstaff, Ariz.: Northland Press, 1987.

Martin, Paul S. "Prehistoric Overkill." In *Pleistocene Extinctions,* Paul S. Martin and H. E. Wright, Jr., eds. New Haven, Conn.: Yale University Press, 1967, pp. 75–120.

———. "Prehistoric Overkill: The Global Model." In *Quaternary Extinctions,* Paul S. Martin and Richard G. Klein, eds. Tucson: University of Arizona Press, 1984, pp. 354–403.

———. "Who or What Destroyed Our Mammoths?" In *Megafauna and Man: Discovery of America's Heartland,* Larry Agenbroad, ed. Flagstaff: Northern Arizona University, 1990, pp. 109–17.

Martin, Paul S., and Richard G. Klein, eds. *Quaternary Extinctions.* Tucson: The University of Arizona Press, 1984.

Martin, Paul S., and H. E. Wright, Jr., eds. *Pleistocene Extinctions.*

New Haven, Conn.: Yale University Press, 1967.

Merriam, John C. "The Fauna of Rancho La Brea. Part I: Occurrence." *Memoirs of the University of Calfornia* [Berkeley] 1, no. 2 (1911): 213.

Montagu, M. F. Ashley. "An Indian Tradition Relating to the Mastodon." *American Anthropologist* 46 (1944): 568–71.

Montagu, M. F. Ashley, and C. Bernard Peterson. "The Earliest Account of the Association of Human Artifacts with Fossil Mammals in North America." *Proceedings of the American Philosophical Society* (1932): 407–19.

Moodie, D. Wayne, A.J.W. Catchpole, and Kerry Abel. "Northern Athapascan Oral Traditions and the White River Volcano." *Ethnohistory* 39, no.2 (spring 1992): 148–71.

Muller, Werner. *America: The New World or the Old?* Frankfurt am Main: Verlag Peter Lang, 1989.

Muller-Beck, H. "Migrations of Hunters on the Land Bridge in the Upper Pleistocene." In *The Bering Land Bridge*, David M. Hopkins, ed. Stanford, Calif.: Stanford University Press, 1967, pp. 373–408.

Neihardt, John. *Black Elk Speaks.* Lincoln: University of Nebraska Press, 1961.

———. *When the Tree Flowered.* New York: Macmillan, 1952.

Nichols, John D., and Ogg, Arden C., eds. *nikotwâsik iskwahtêm, pâskihtêpayih! Studies in Honour of H.C. Wolfart.* Algonquian and Iroquoian Linguistics, Memoir 13, 1996.

Patten, Donald. "A Comprehensive Theory on Aging, Gigantism and Longevity." *Catastrophism and Ancient History* 2, pt. 1 (August 1979): 13–60.

———. *The Biblical Flood and the Ice Epoch.* Seattle, Wash.: Pacific Meridian Publishing, 1966.

Pendergast, David M., and Clement W. Meighan. "Folk Traditions as Historical Fact: A Paiute Example." *Journal of American Folklore* 72 (1959): 128–33.

Raglan, Lord. "Folk Traditions as Historical Fact." Notes and Queries, *Journal of American Folklore* 73 (1960): 59–60.

Siebert, F. T., Jr. Discussion and Correspondence: "Mammoth or 'Stiff-Legged Bear.'" *American Anthropologist* 39 (1937): 721–25.

Simpson, George Gaylord. "Mammals and Land Bridges." *Journal of the Washington Academy of Sciences* 30, no. 4 (April 1940): 137–63.

———. "The Discovery of Fossil Vertebrates in North America." *Journal of Paleontology* 17, no. 1 (1943): 26–38.

Spielmann, Peter. "Botanists Giddy Over Rare Find in Australia." *The Denver Post,* December 15, 1994, p. 11A.

Standing Bear, Luther. *Land of the Spotted Eagle.* Boston: Houghton Mifflin, 1933.

Steward, Julian. "Some Western Shoshoni Myths." Washington, D.C.: Smithsonian Institution, Bureau of American Ethnology, Bulletin 136, Anthropological Papers, no. 31, 1943.

Stock, Chester. *Rancho La Brea: A Record of Pleistocene Life in California,* 7th ed. Los Angeles: Natural History Museum, Science Series no. 37, 1992.

Strong, W. D. "North American Indian Traditions Suggesting a Knowledge of the Mammoth." *American Anthropologist* 36 (1934): 81–88.

Swan, James G. *The Indians of Cape Flattery.* Washington, D.C.: Smithsonian Contributions to Knowledge, no. 220, 1869.

Taber, Stephen. "Perennially Frozen Ground in Alaska: Its Origins and History." *Geological Society of America Bulletin* 54 (1943): 1,433–1,548.

Taylor, Gordon Rattray. *The Great Evolution Mystery.* New York: Harper & Row, 1983.

Toynbee, Arnold. *A Study of History.* New York: Oxford University Press, 1947.

Tyon, Thomas. "Spirits." In *James Walker, Lakota Belief and Ritual,* Raymond De Malle, ed. Lincoln: University of Nebraska Press, 1980.

Vangengeim, E. A. "Quaternary Mammalian Faunas of Siberia and North America." In *The Bering Land Bridge,* David M. Hopkins, ed. Stanford, Calif.: Stanford University Press, 1967, pp. 281–87.

Velikovsky, Immanuel. *Worlds in Collision.* New York: Macmillan, 1950.

———. *Earth in Upheaval.* New York: Doubleday, 1955.

Vereshchagin, N. K. "Primitive Hunters and Pleistocene Extinction in the Soviet Union." In *Pleistocene Extinctions,* Paul S. Martin and H. E. Wright, Jr., eds. New Haven, Conn.: Yale University Press, 1967, pp. 365–98.

Vitaliano, Dorothy. "Geomythology." *Indiana University Folklore Institute Journal* 5 (1968): 5–30.

———. *Legends of the Earth.* Bloomington: Indiana University Press, 1973.

Walker, Deward. "Tribal Account of Post-Pleistocene Floods from Ice Dams in the Plateau." Unpublished manuscript on Spokane flood story, August 1994.

Walker, James. *Lakota Belief and Ritual,* Raymond DeMallie, ed. Lincoln: University of Nebraska Press, 1980.

Warshofsky, Fred. *Doomsday: The Science of Catastrophism.* New York: Readers Digest Press, 1977.

Waters, Frank. *The Book of the Hopi.* New York: Penguin Books, 1963.

Wedel, Waldo R. "The Great Plains." In *Prehistoric Man in the New World,* Jesse D. Jennings and Edward Norbeck, eds. Chicago: University of Chicago Press, 1963, pp. 193–220.

Whitehead, Alfred North. *Adventures of Ideas.* New York: Macmillan, 1933.

Whitley, Reverend D. Gath. "The Ivory Islands in the Arctic Ocean." *Philosophical Society of Great Britain* 12 (1910): 35–56.

Woodard, Geoffrey, and Leslie Marcus. "Rancho La Brea Fossil Deposits: A Re-evaluation from Stratigraphic and Geological Evidence." *Journal of Paleontology* 47, no. 1 (January 1973): 54–69.

Worster, Donald. *Dust Bowl.* New York: Oxford University Press, 1979.

Wrangell, Admiral Ferdinand. *Narrative of an Expedition to the Polar Sea in the Years 1820, 1821, 1822, and 1823 Commanded by Lieutenant, now Admiral Ferdinand Wrangell of the Russian Imperial Navy.* New York: Harper & Brothers, 1841.

Wright, G. Frederick. *The Ice Age in North America.* New York: D. Appleton, 1890.

Index